ACE ORGANIC CHEMISTRY I

(THE EASY GUIDE TO ACE ORGANIC CHEMISTRY I)

BY: DR. HOLDEN HEMSWORTH

Copyright © 2015 by Holden Hemsworth

All rights reserved. No part of this publication may be reproduced, distributed, or transmitted in any form or by any means, including photocopying, recording, or other electronic or mechanical methods, without the prior written permission of the publisher, except in the case of brief quotations embodied in critical reviews and certain other noncommercial uses permitted by copyright law.

DISCLAIMER

Chemistry, like any field of science, is continuously changing and new information continues to be discovered. The author and publisher have reviewed all information in this book with resources believed to be reliable and accurate and have made every effort to provide information that is up to date and correct at the time of publication. Despite our best efforts we cannot guarantee that the information contained herein is complete or fully accurate due to the possibility of the discovery of contradictory information in the future and any human error on part of the author, publisher, and any other party involved in the production of this work. The author, publisher, and all other parties involved in this work disclaim all responsibility from any errors contained within this work and from any results that arise from the use of this information. Readers are encouraged to check all information in this book with institutional guidelines, other sources, and up to date information.

MCAT® is a registered trademark of the Association of American Medical Colleges and holds no affiliation with this book.

The information contained in this book is provided for general information purposes only and does not constitute medical, legal or other professional advice on any subject matter. The author or publisher of this book does not accept any responsibility for any loss which may arise from reliance on information contained within this book or on any associated websites or blogs.

Why I Created This Study Guide

Organic Chemistry is typically taught over two semesters in college and these courses tend to be some of the hardest for students as they require a lot of memorization. In this book, I try to breakdown the content covered in the typical first semester of an Organic Chemistry course for easy understanding and to point out the most important subject matter that students are likely to encounter in hopes of making the material more palatable. This book is meant to be a supplemental resource to lecture notes and textbooks, to boost your learning, and to go hand in hand with your studying!

I am committed to providing my readers with books that contain concise and accurate information and I am committed to providing them tremendous value for their time and money.

Best regards,

Dr. Holden Hemsworth

Table of Contents

CHAPTER 1: Revisiting General Chemistry 1

CHAPTER 2: Alkanes and Cycloalkanes 17

CHAPTER 3: Stereoisomerism and Chirality 30

CHAPTER 4: Acids and Bases 39

CHAPTER 5: Alkenes 44

CHAPTER 6: Reactions of Alkenes 52

CHAPTER 7: Alkynes and Reactions of Alkynes 75

CHAPTER 8: Haloalkanes and Radical Reactions 88

CHAPTER 9: Nucleophilic Substitution and β-elimination 96

CHAPTER 10: Alcohols and their Reactions 104

CHAPTER 11: Ethers and Epoxides 113

CHAPTER 1: REVISITING GENERAL CHEMISTRY

Organic Chemistry

Organic chemistry is the branch of chemistry that specializes in studying carbon compounds. Organic compounds contain both carbon and hydrogen atoms, while inorganic compounds typically lack carbon.

Carbon

- Relatively small atom
- Capable of forming single, double, and triple bonds
- Electronegativity = 2.55
    - Intermediate electronegativity
- Forms strong bonds with C (carbon), H (hydrogen), O (oxygen), N (nitrogen)
    - Also with some metals
- Has 4 valence electrons
    - To fill its outer shell, it typically forms four covalent bonds
    - Carbon is capable of making large and complex molecules because it is capable of branching off into four directions
- Covalent bonds link carbon atoms together into long chains
    - Form the skeletal framework for organic molecules
- Hydrocarbons are molecules containing only carbon and hydrogen
    - Examples: methane (CH_4), ethane (C_2H_6), propane (C_3H_8)
    - Hydrocarbon chains are hydrophobic because they consist of nonpolar bonds

Electron Orbitals

Electrons orbit the nucleus of an atom in "orbitals" of increasing energy levels, or shells. Orbitals are mathematical functions that describe the wave-like behavior of an electron in a molecule (calculates the probability of where you might find an electron).

- Electrons in shells closest to the nucleus have the lowest potential energy
    - Conversely, shells farther from the nucleus have higher potential energy

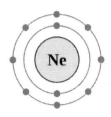

Shell Model of a Neon Atom:

- Orbitals aren't necessarily circular as represented in the shell model
    - In reality, orbitals are "clouds" of various shapes
    - Each orbital can only hold a limited number of electrons
    - An atom can have multiple orbitals of different shapes
- Electrons may move from one energy level to another
    - Happens when they gain or lose energy equal to the difference in potential energy between energy levels
- First energy level:
    - One spherical s orbital (1s orbital)
    - Holds up to two electrons
- Second energy level
    - One spherical s orbital (2s orbital)
    - Three dumbbell-shaped p orbitals ($2p_x$, $2p_y$, $2p_z$ orbitals)
- Higher energy levels
    - Contain s and p orbitals
    - Contain other orbitals with more complex shapes

Orbital Shapes (s, p, d, f) Top to Bottom:

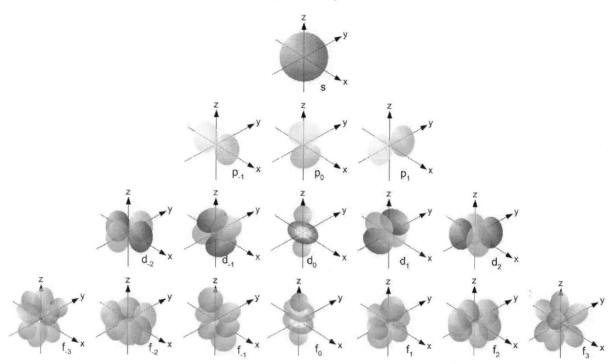

Electron Configuration

The electron configuration of an atom refers to the particular distribution of electrons among the available sub shells in that atom.

- Electronic configuration notation lists subshell symbols (s, p, d, f) sequentially with a superscript to indicate the number of electrons in that subshell
    - Example: Carbon
        - Atomic Number: 6
        - Number of electrons in a neutral carbon atom: 6
            - Number of electrons for a neutral atom is the same as its atomic number
        - 2 electrons in the "1s" sub shell
        - 2 electrons in the "2s" sub shell
        - 2 electrons in the "2p" sub shell
        - Electron Configuration: $1s^2 2s^2 2p^2$

- Configurations can become quite complex as atomic number increases
    - To remedy this, a condensed form of the configuration is often used which utilizes electron configurations of noble gases
        - Noble gases have the maximum number of electrons possible in their outer shell
            - Makes them very unreactive
        - The noble gases are: Helium, Neon, Argon, Krypton, Xenon, and Radon

Table of Condensed Electronic Configuration Examples:

Element	Noble Gas?	Full Electronic Configuration	Condensed Electronic Configuration	Total Number of Electrons
Neon	Yes	$1s^2 2s^2 2p^6$	[Ne]	10
Argon	Yes	$1s^2 2s^2 2p^6 3s^2 3p^6$	[Ar]	18
Krypton	Yes	$1s^2 2s^2 2p^6 3s^2 3p^6 3d^{10} 4s^2 4p^6$	[Kr]	36
Beryllium	No	$1s^2 2s^2$	[He] $2s^2$	4
Magnesium	No	$1s^2 2s^2 2p^6 3s^2$	[Ne] $3s^2$	12
Calcium	No	$1s^2 2s^2 2p^6 3s^2 3p^6 4s^2$	[Ar] $4s^2$	20

   - [X] represents the electron configuration of the nearest noble gas that appears before the element of interest on the periodic table
- Keep in mind that you have to adjust the number of electrons and thus the electron configuration for cations and anions of an element

Energy-level Diagrams

- Energy-level diagrams are notations used to show how the orbitals of a sub shell are occupied by electrons
    - Each group of orbitals is labeled by its sub shell notation (s, p, d, f)
    - Electrons are represented by arrows

Energy-level Diagram for Carbon:

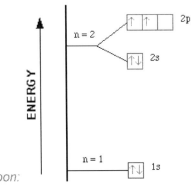

Lewis Dot Structures

Lewis Dot Structure of Carbon:

- Symbol of the element represents the nucleus and all the electrons in the inner shells
    - Dots represent electrons in the valence shell
        - Valence shell – outermost electron shell of an atom that is occupied with electrons
        - Valence electrons – electrons in the valence shell
            - These are the electrons primarily involved in chemical bonding and chemical reactions
    - Bonding electron pairs are represented by either two dots or a dash

Lewis Electron-dot Formula Example:

$$H \cdot + :\ddot{Cl} \cdot \longrightarrow H-\ddot{Cl}: \text{ or } H:\ddot{Cl}:$$

- Rules for Forming Lewis Structures
    - Calculate the number of valence electrons for the molecule
        - Group # for each atom (1-8)
            - Gives valence electron number for each atom
            - Add all numbers up
        - Add the charge of any anions
            - Example: an anion with a -2 charge has 2 extra electrons, you would add 2 to the total count
        - Subtract the charge of any cations
            - Example: a cation with a +3 charge lacks 3 electrons, you would subtract 3 from the total count
    - Place the atom with the lowest group number and lowest electronegativity as the central atom
    - Arrange the other elements around the central atom

- Distribute electrons to atoms surrounding the central atom to satisfy the octet rule for each atom
- Distribute the remaining electrons as pairs to the central atom
- If the central atom is deficient in electrons, complete the octet for it by forming double bonds or possibly a triple bond

Ions

Ions are charged atoms or molecules. Ions are formed when atoms or groups of atoms gain or lose valence electrons.

- Monatomic ion – single atom with more or less electrons than the number of electrons in the atom's neutral state
- Polyatomic ions – group of atoms with excess or deficient number of electrons
- Anion – negatively charged ion
- Cation – positively charged ion
- Ionic compounds – association of a cation and an ani//on

Electronegativity and Ions

Electronegativity is the measure of an atom's ability of to draw bonding electrons to itself in a molecule.

- Electronegativity tends to increase from the lower-left corner to the upper-right corner of the periodic table

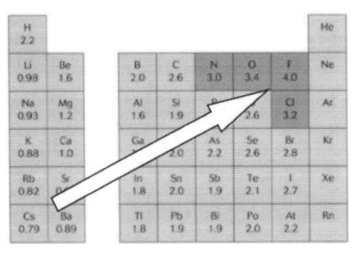

Electronegativity Trend:

Types of Bonds

Covalent Bonds

- Two atoms share valence electrons
- Indicates that atomic orbitals are overlapping
    - Overlapping requires proximity and orientation
- Two Types
    - Non-polar covalent bond – electrons shared equally between atoms
        - Electronegativity of the two atoms is about the same
        - Typically electronegativity difference between the two atoms has to be less than 0.5 for non-polar bonds
        - Electronegativity – an atom's ability to attract and hold on to electrons, represented by a number
    - Polar covalent bonds – electrons shared disproportionately between atoms
        - Electronegativity between the two atoms is different by a greater degree than 0.5 but less than 2.0
        - Polarity can be represented using δ+ and δ-
            - δ+ represents the positive end
            - δ- represents the negative end

$$\overset{\delta+ \quad \delta-}{H\text{—}Cl}$$

- Polarity can also be represented by an arrow with a plus sign tail
    - Tip of the arrow represents the negative end
    - Plus sign tail represents the positive end

- Number of shared pairs
    - Single bond - one shared pair
    - Double bond – two shared pair
    - Triple bond – three shared pairs

Ionic Bonds

- Electrons are transferred, not shared between atoms
- An atom with high electronegativity will take an electron from an atom with low electronegativity
    - Typically, difference in electronegativity is more than 2.0
- Ion – charged atom or molecule
    - Anion – negatively charged ion
    - Cation – positively charged ion

Hydrogen Bonds

- Attractive force between a hydrogen attached to an electronegative atom of one molecule to a hydrogen attached to an electronegative atom of a different molecule
- Electronegative atoms usually seen in molecules are O, N, and F

Van der Waals Forces

A general term used for the attraction of intermolecular forces between molecules.

Dipole-dipole Interactions

- Interaction between 2 polar groups

London Dispersion Forces

- Interaction between 2 non-polar molecules
- Small fluctuation in electronic distribution

Intermolecular Forces

Forces that act between neighboring particles (can be repulsive or attractive).

- Intermolecular bond strength ranking (strong to weak):
    - Covalent > ionic > hydrogen > van der Waals forces

- Weaker bonds and forces are easily broken or overcome and also re-formed
    - Makes them vital for the molecular dynamics of life
    - Shared electron pair simultaneously fills the outer level of both atoms

Functional Groups

Functional groups are characteristic groups of atoms responsible for the characteristic reactions of a particular compound.

- Functional groups have specific chemical and physical properties that are associated with them
- Are commonly the chemically reactive regions within organic compounds
    - Determine unique chemical properties of organic molecules that they are a part of
    - Consistent properties in all compounds in which they occur

Common Functional Groups

- Hydroxyl group - consist of a hydrogen atom bonded to an oxygen atom

    Hydroxyl Group: O–H

    - Polar group; oxygen and hydrogen bond is a polar covalent bond
    - Organic compounds with hydroxyl groups are called alcohols
    - Alcohol classification
        - Primary (1°) – 1 carbon atom bonded to the carbon bearing the hydroxyl group
        - Secondary (2°) - 2 carbon atoms bonded to the carbon bearing the hydroxyl group
        - Tertiary (3°) - 3 carbon atoms bonded to the carbon bearing the hydroxyl group

$$CH_3-\underset{H}{\overset{H}{C}}-OH \qquad CH_3-\underset{CH_3}{\overset{H}{C}}-OH \qquad CH_3-\underset{CH_3}{\overset{CH_3}{C}}-OH$$

Primary (1°) Alcohol   Secondary (2°) Alcohol   Tertiary (3°) Alcohol

- Amino group - consists of a nitrogen atom bonded to two hydrogens and to the carbon skeleton

  Amino Group:

  - Amines – consist of an amino group bonded to either one, two, or three carbons (1°, 2°, or 3°)

  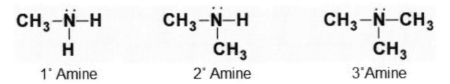

- Carbonyl group - consists of a carbon atom double-bonded to oxygen

  Carbonyl Group: $-\overset{\overset{O}{\|}}{C}-$

- Aldehyde – carbonyl group with a hydrogen attached to the carbon

  Aldehyde Group:

- Ketone – carbonyl group with no hydrogens attached to the carbon

- Carboxyl group – consists of a carbon atom which is attached by a double-bond to an oxygen and single-bonded to the oxygen of a hydroxyl group

  - Group has acidic properties
    - Since it donates H^+
  - Organic compounds with a carboxyl group are called **carboxylic acids**

  Carboxyl Group:

- Ester – derivative of carboxylic acid, where the hydrogen bond is replaced with a carbon bond

- Amide (aka carboxylic acid) – derivative of carboxylic acid in which the hydroxyl group (-OH) is replaced by an amine

  Amide:

- Sulfhydryl group - consists of an sulfur atom bonded to a hydrogen
    - Organic compounds with a sulfhydryl group are called thiols

  Sulfhydryl Group:

- Phosphate group – consists of a phosphorous atom single bonded to 4 oxygen atoms, and one of those oxygens is attached to the rest of the molecule
    - Acidic properties (loses H^+)
    - Organic phosphates are important part of cellular energy storage and transfer

  Phosphate Group:

Molecular Orbital Theory

- As atoms approach each other and their atomic orbitals overlap, molecular orbitals are formed
    - Only outer (valence) atomic orbitals interact enough to form molecular orbitals
- Combining atomic orbitals to form molecular orbitals involves adding or subtracting atomic wave functions
- Adding wave functions
    - Forms a bonding (σ) molecular orbital
    - Electron charge between nuclei is dispersed over a larger area than in atomic orbitals

- Molecular orbitals have lower energy than atomic orbitals
    - Reduction in electron repulsion
- Bonding molecular orbital is more stable than atomic orbital

- Subtracting Wave Functions
    - Forms an antibonding (σ^*) molecular orbital
    - Electrons do not shield one nuclei from the other
        - Results in increased nucleus-nucleus repulsion
    - Antibonding molecular orbitals have a higher energy than the corresponding atom orbitals
    - When the antibonding orbital is occupied, the molecule is less stable than when the orbital is not occupied

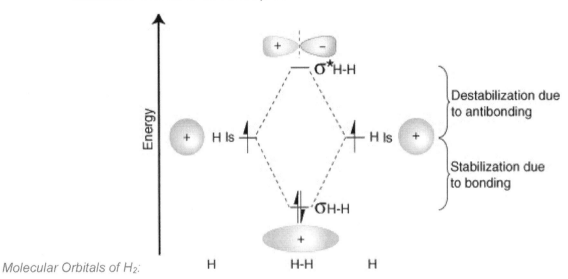

Molecular Orbitals of H_2:

Hybrid Orbitals

- Quantum mechanical calculations show that if specific combinations of orbitals are mixed, "new" atomic orbitals are formed
    - These new orbitals are called hybrid orbitals
- Types of hybrid orbitals
    - Each type has a unique geometric arrangement

Hybrid Orbitals (Hybridization)	Geometric Arrangements	Number of Hybrid Orbitals Formed by Central Atom	Example
sp	Linear	2	Be in BeF_2
sp^2	Trigonal planar	3	B in BF_3
sp^3	Tetrahedral	4	C in CH_4
sp^3d	Trigonal bipyramidal	5	P in PCl_5
sp^3d^2	Octahedral	6	S in SF_6

- Hybrid orbitals are used to describe bonding that is obtained by taking combinations of atomic orbitals of an isolated atom

- Number of hybrid orbitals formed = number of atomic orbitals combined

- Steps for determining bonding description

    o Write the Lewis dot formula for the molecule

    o Then use the VSEPR theory to determine the arrangement of electron pairs around the central atom

    o From the geometric arrangement, determine the hybridization type

    o Assign valence electrons to the hybrid orbitals of the central atom one at a time

        ▪ Pair only when necessary

    o Form bonds to the central atom by overlapping singly occupied orbitals of other atoms with the singly occupied hybrid orbitals of the central atom

Multiple Bonds

- Orbitals can overlap in two ways

    o Side to side

    o End to end

- Two types of covalent bonds
    - Sigma bonds (C-C)
        - Formed from an overlap of one end of the orbital to the end of another orbital
    - pi bonds (C=C)
        - Formed when orbitals overlap side to side
        - Creates two regions of electron density
            - One above and one below
- Double bonds always consist of one sigma bond and one pi bond

Covalent Bonding of Carbon

Groups Bonded to Carbon	Orbital Hybridization	Predicted Bond Angles	Types of Bonds to Each Carbon
4	sp^3	109.5°	Four σ bonds
3	sp^2	120°	Three σ bonds One π bond
2	sp	180°	Two σ bonds Two π bonds

Resonance (Delocalized Bonding)

- Structures of some molecules can be represented by more than one Lewis dot formula
    - Individual Lewis structures are called contributing structures
    - Individual contributing structures are connected by double-headed arrows (aka resonance arrows)
    - Molecule or ion is a hybrid of the contributing structures and displays delocalized bonding
        - Delocalized bonding is where a bonding pair of electrons is spread over a number of atoms
- Some resonance structures contribute more to the overall structure than others
    - How to determine which structures are more contributing:

- Structures where all atoms have filled valence shells
- Structures with the greater number of covalent bonds
- Structures with less charges
  - Formal charges can help discern which structure is most likely (discussed later in this section)
- Structures that carry a negative charge on the more electronegative atom

Example of Resonance Structures:

- Curved arrow – symbol used to the redistribution of valence electrons
  - Always drawn as noted in the figure below

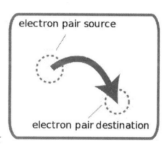

How Curved Arrows are Drawn:

Formal Charge

- An atom's formal charge is:
  - Total number of valence electrons
  - Minus all unshared electron
  - Minus ½ of its shared electrons

- Formal charges have to sum to the actual charge of the species
  - 0 charge for a neutral molecule
  - Ionic charge for an ion
- Lewis structures with the smallest formal charge are the most likely to occur

Formal Charge vs. Oxidation Number

- Formal charges are used to examine resonance hybrid structures
  - Oxidation numbers are used to monitor redox reactions
- **Formal Charge**
  - Bonding electrons are assigned equally to the atoms
    - Each atom has half the electrons making up the bond
  - Formal Charge = valence e^- – (unbonded e^- + ½ bonding e^-)
- **Oxidation Number**
  - Bonding electrons are transferred completely to the more electronegative atom
  - Oxidation Number = valence e^- – (unbonded e^- + bonding e^-)

Chapter 2: Alkanes and Cycloalkanes

Terminology

- Hydrocarbons - molecules containing only carbon and hydrogen
    - Examples: methane (CH_4), ethane (C_2H_6), propane (C_3H_8)
- Saturated hydrocarbon – hydrocarbon containing only single bonds
- Unsaturated hydrocarbon – hydrocarbon containing at least one double bond
- Alkane (aka aliphatic hydrocarbon) – saturated hydrocarbon whose carbons are arranged in an open chain
    - General formula: C_nH_{2n+2}
- Cycloalkanes – hydrocarbon with a ring of carbon atoms joined by single bonds

Classification of Carbon and Hydrogen

- Primary (1°) Carbon - carbon bonded to one other carbon
    - 1° H - hydrogen bonded to a 1° carbon
- Secondary (2°) Carbon - carbon bonded to two other carbons
    - 2° H - hydrogen bonded to a 2° carbon
- Tertiary (3°) Carbon - carbon bonded to three other carbons
    - 3° H - hydrogen bonded to a 3° carbon
- Quaternary (4°) Carbon - a carbon bonded to four other carbons

Drawing Alkanes

- Line-angle formulas - abbreviated way of drawing structural formulas
    - Each line represents a C-C bond
    - Each vertex and line ending represents a carbon

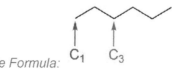

Example of a Line-angle Formula:

- C_1 is a carbon represented by the end of a line

- C_3 is a carbon represented by a vertex
  - Hydrogens are not shown, they are assumed to be there
    - C_1 in the example above has 3 hydrogens and is bonded to C_2 (4 total bonds)
    - C_3 in the example above has 2 hydrogens, its bonded to C_2 and C_3 (4 total bonds)
  - Elements aside from hydrogen and carbon are always shown

IUPAC Nomenclature

IUPAC (International Union of Pure and Applied Chemistry) nomenclature is a systematic method of naming organic chemical compounds.

IUPAC - General

- Parent chain – longest carbon chain in a molecule
  - The parent name is used to specify the number of carbon atoms in the parent chain

Parent Name	Number of Carbons	Parent Name	Number of Carbons
Meth-	1	Undec-	11
Eth-	2	Dodec-	12
Prop-	3	Tridec-	13
But-	4	Tetradec-	14
Pent-	5	Pentadec-	15
Hex-	6	Hexadec-	16
Hept-	7	Heptadec-	17
Oct-	8	Octadec-	18
Non-	9	Nonadec-	19
Dec-	10	Eicos-	20

- Infix is used to inform about the type of Carbon-Carbon bonds in the parent chain

Infix	Carbon-Carbon Bonds in the Parent Chain
-an-	All single bonds
-en-	One or more double bonds
-yn-	One or more triple bonds

- Suffix is used to inform about the class of compound

Suffix	Class
-e	Hydrocarbon
-ol	Alcohol
-al	Aldehyde
-amine	Amine
-one	Ketone
-oic acid	Carboxylic Acid

- Substituent – group bonded to the parent chain
    - Alkyl group – substituent derived by removal of a hydrogen from an alkane
        - Alkyl groups are symbolized by the capital letter "R"

Common Alkyl Group Substituents:

Alkyl Group	Substituent Name
CH_3-	Methyl
CH_3CH_2-	Ethyl
$CH_3CH_2CH_2-$	Propyl
$-CHCH_3$ | CH_3	Isopropyl
$CH_3CH_2CH_2CH_2-$	Butyl
$-CH_2CHCH_3$ | CH_3	Isobutyl
$-CHCH_2CH_3$ | CH_3	*Sec*-butyl
CH_3 | $-CCH_3$ | CH_3	*Tert*-butyl

Naming Alkanes

- Suffix –ane specifies an alkane (e.g., eth<u>ane</u>, meth<u>ane</u>)
- Identify the parent chain (longest Carbon chain) and number it (always number sequentially)

- Example: [structure with carbons numbered 6-5-4-3-2-1]

    - If there are no substituents, as in the example above, you can begin numbering from either end

  o Number of carbons in the parent chain gives you the parent name, then add the suffix –ane

    - In the example above, there are 6 carbons so the parent name is hex- and you would add the suffix –ane to get "hexane"

- Each substituent has a name and a number (use a hyphen to connect the name and number)

  o Number of the substituent is determined by which carbon it is on

  o Examples:

    [structure] Name: 2-methylbutane

    - Methyl group (CH_3-) is on C_2 so it is named 2-methyl

    [structure] Name: 3-methylpropane

    - Methyl group is on C_3 so it is named 3-methyl

- Numbering the parent chain must be done so that substituents get the smallest possible numbers

  o Examples:

    correct
    [structure numbered 6-5-4-3-2-1 with CH₃]

    incorrect
    [structure numbered 1-2-3-4-5-6 with CH₃]

    Correct Name: 2-methylhexane

correct

[Structure: CH₃ groups on C2 and C4 of a 6-carbon chain numbered 1-6]

Incorrect:

[Same structure numbered 6-1 in reverse]

Correct Name: 2,4- dimethylhexane

- If there are two or more of the same substituent, add a comma to separate the substituent numbers and add a prefix to indicate how many of the substituents you have

    - Two of the same substituent (di-)
    - Three of the same substituent (tri-), and so on and so forth

- If there are two or more different substituents
    - List them in alphabetical order
    - Example:

    [Structure with CH₃ on C2 and CH₂CH₃ on C4 of an 8-carbon chain]

    Name: 4-ethyl-2-methyloctane

    - Prefixes (e.g., di-, tri-) are not included in alphabetization
        - Example:

    [Structure with H₃C and CH₃ on C2, and CH₂CH₃ on C4 of a 6-carbon chain]

    Name: 4-ethyl-2,2-dimethylhexane

Cycloalkanes

- General formula: C_nH_{2n}
    - Five and six-membered rings are the most common

- Structure and nomenclature
    - Add prefix cyclo- to the name of the open-chain alkane containing the same number of carbons
        - Example: six carbon open-chain alkane with no substituents would be called a "hexane," a six carbon ring would be called a cyclohexane

    - If there is only one substituent in the ring structure, it does not need to be assigned a number
    - If there are two substituents, start numbering from the substituent that comes first alphabetically
    - If there are three or more substituents, number the ring so that the substituents have the lowest possible set of numbers

Newman Projection

- Newman projection – way of visualizing chemical conformations of a carbon-carbon bond
    - Conformations - any 3D arrangements of atoms in a molecule that result from rotation around a single bond

Example of Newman Projections

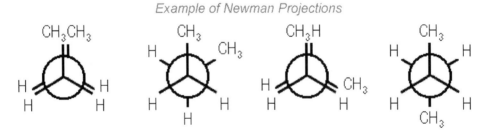

- Newman projection conventions
    - Chemical bond is viewed from front to back
        - Front carbon represented by a dot
        - Back carbon represented by a circle
        - Bonds represented by straight lines

- Staggered conformation – atoms or groups on one carbon are as far apart as possible from the atoms or groups on an adjacent carbon
    - Two types
        - Anti – conformation about a single bond in which the groups on adjacent carbons lie at a dihedral angle of 180°
            - Dihedral angle (θ) - angle between two bonds originating from different atoms in a Newman projection

Staggered (Anti) Conformation of Butane:

- Gauche – conformation about a single bond in which two groups on adjacent carbons lie at a dihedral angle of 60°

Staggered (Gauche) Conformation of Butane: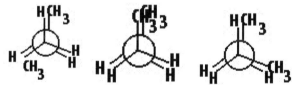

- Eclipsed conformation - atoms or groups of atoms on one carbon are as close as possible to the atoms or groups of atoms on an adjacent carbon

Eclipsed Conformations of Butane:

Strain and Energy

Strain energy is the increase in energy that results from the distortion of bond angles and bond lengths from their optimal values.

- Steric strain (aka nonbonded interaction strain) – increases in potential energy of a molecule due to repulsion between electrons in atoms that are not directly bonded to each other

Highest Steric Strain Conformation of Butane: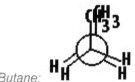

- o Conformation of butane shown above has the highest steric strain out of all the other conformation, since the "bulky" methyl group (-CH3) are closest together in this conformation
- Angle strain – increase in potential energy due to bond angles deviating from their optimal value
- Torsional strain - strain that emerges when non-bonded atoms separated by three bonds are forced from a staggered conformation into an eclipsed conformation

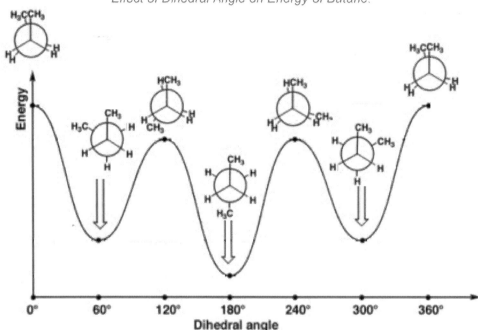

Effect of Dihedral Angle on Energy of Butane:

Conformations of Cyclohexane

Flat drawings do not accurately represent the actual 3D shape of a five- or six-membered ring.

Chair Conformation

- Chair conformation – most stable puckered conformation of a cyclohexane ring
  - Most stable conformation that minimizes strain
    - Bond angles are 110.9°
      - Ideal bond angle
    - Bonds on all adjacent carbons are staggered

Cyclohexane Flat (Left) and Chair (Right) Conformations:

○ Six of the hydrogens are "axial" and six of them are equatorial

- Axial hydrogens – hydrogens that are parallel to the axis of the ring

  • Axial bonds are always drawn straight up or straight down

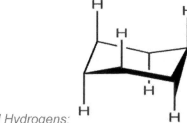

Axial Hydrogens:

- Equatorial Hydrogens – Hydrogens that are perpendicular to the axis of the ring

  • Equatorial bonds are always drawn parallel to the lines representing the C-C bonds in the ring

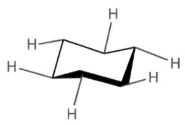

Equatorial Hydrogens:

- There are two chair conformations, one chair conformation can be used to determine the other chair conformation through the process of "flipping the chair" (aka ring flip)

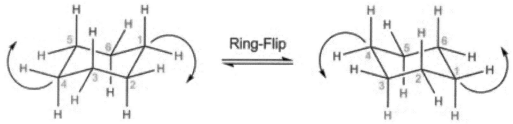

Ring Flip:

- When the chair is flipped, all axial positions become equatorial
  - All equatorial positions become axial
- ***IMPORTANT*:** Substituents are more stable in the equatorial position
  - Equatorial position is preferred because there is unfavorable steric interactions that occur between axial atoms on the same side
    - This unfavorable interaction is called 1,3–diaxial interaction

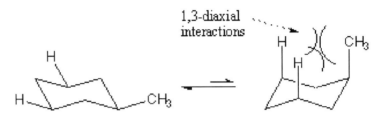

1,3-diaxial Interaction: **equatorial methyl**   **axial methyl**

Two Conformations of Methylcyclohexane:

- Conformation for methylcyclohexane on the right is more stable, since the methyl group (-CH₃) is in the equatorial position

Boat Conformation

- Boat conformation – a puckered conformation of a cyclohexane ring where carbons 1 and 4 are bent towards one another

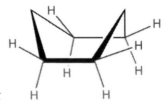

Cyclohexane Boat Conformation:

- Flagpole hydrogens – hydrogens in a 1,4 – relationship in a boat cyclohexane

Flagpole Hydrogens:

  - Boat conformation is less stable than chair conformation because the groups involved in the 1,4 relationship create steric strain

- Steric hindrance can be partially relieved with the twist boat conformation
  - Twist boat conformation is still less stable than the chair conformation

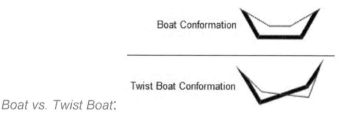

Boat vs. Twist Boat:

Cis, Trans Isomers

Stereoisomers are compounds that have the same molecular formula, same connectivity, but a different orientation of their atoms in space. Cis and trans isomers are stereoisomers that result from either a ring or a double bond.

- Stereocenter – an atom bearing groups where exchange of two groups produces a different stereoisomer
- Configuration – refers to the arrangement of atoms about a stereocenter
- The following are not isomers:

  - They are not isomers because there is free-rotation around single bonds

- The following are isomers:

  [structures of 1,2-dichloroethene isomers]

    - They are isomers because there is no free-rotation around double bonds
        - Chlorine atoms are locked in their positions

- *Trans* (latin meaning "across") isomers – functional groups are on opposite sides

  trans-1,2-dichloroethene:

- *Cis* (latin meaning "on this side") isomers – functional groups are on the same side

  cis-1,2-dichloroethene:

Solid and Dashed Wedges

- Solid wedge - symbol used to indicate that a bond is projecting out towards the person viewing the bond

- Dashed wedge – symbol used to indicate that a bond or group is pointed away from the person viewing the bond

- Sold line – symbol used to indicate that the bond lies in the plane of the paper

  1,4-Dimethylcyclohexane:

  ***trans* -1,4-Dimethyl-cyclohexane**   ***cis*-1,4-Dimethyl-cyclohexane**

Physical Properties of Alkanes

- Not very reactive
- Little biological activity
- Colorless
- Odorless
- Low molecular weight alkanes are gases at room temperature
    - E.g., methane and butane
- Intermediate molecular weight alkanes are liquids at room temperature
- High molecular weight alkanes are solid at room temperature
- Insoluble in water
    - But, dissolve in organic solvents
- Trends
    - Linear alkanes have higher melting/boiling points than their branched counterparts
        - Due to better stacking and surface area contact
    - Highly branched alkanes have **higher** melting points than slightly branched alkanes
        - Due to better stacking
    - Highly branched alkanes have a **lower** boiling point than slightly branched alkanes
        - Due to highly branched alkanes having less surface area

Chapter 3: Stereoisomerism and Chirality

Isomers

Isomers are compounds that have the same molecular formula but vary in their structures and properties.

Types of Isomers

- Constitutional isomers (aka structural isomers - differ in the covalent arrangement of their atoms)

  Structural Isomers: Ethanol (Alcohol)　　　Dimethylether

- Stereoisomers (aka spatial isomers) - share the same covalent bonds, but differ in their spatial arrangements

  o Two types of stereoisomers

    ▪ Enantiomers – two compounds that are mirror images of each other

      • Can occur when four different atoms or groups of atoms are bonded to the same carbon (chiral carbon/asymmetric carbon)

      • Usually one form of an enantiomer is biologically active while the other is not

      Enantiomers: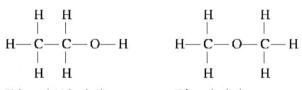

    ▪ Diastereomers – two compounds that are not mirror images of each other

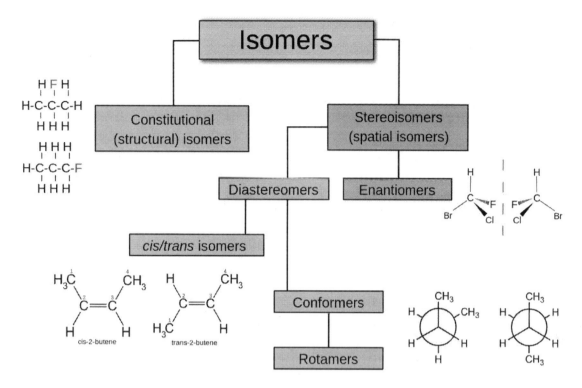

Chirality

- Chiral – an object that is asymmetric in such a way that it is not superposable on its mirror image

- Achiral – an object that lacks chirality

    o Achiral objects have at least one element of symmetry (a plane of symmetry or a center of symmetry)

        ▪ Plane of symmetry – imaginary plane through an object that divides the object into two halves which are mirror images of one another

        ▪ Center of symmetry – any point situated in such a manner that identical components of the object are located on opposite sides and equidistant from that point along any axis that passes through that point

- Carbon bonded to four different groups is the most common cause of chirality in organic molecules

    o All chiral centers are stereocenters

        ▪ Not all stereocenters are chiral centers

- For a molecule with *n* chiral centers, the maximum number of stereoisomers is determined by 2^n
    - Examples
        - Molecule with 1 chiral center, 2^1 or 2, stereoisomers are possible
        - Molecule with 2 chiral centers, 2^2 or 4, stereoisomers are possible
- Meso compound – achiral compound with multiple stereoisomers that is superimposable on its mirror image
    - Meso compounds have an internal plane of symmetry
    - Example: consider tartaric acid

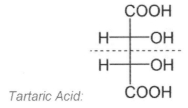

Tartaric Acid:

- Has 2 chiral centers, 2^2 or 4, stereoisomers are predicted but really there are only 3

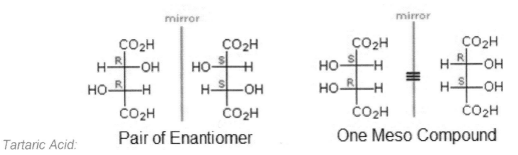

Tartaric Acid: Pair of Enantiomer   One Meso Compound

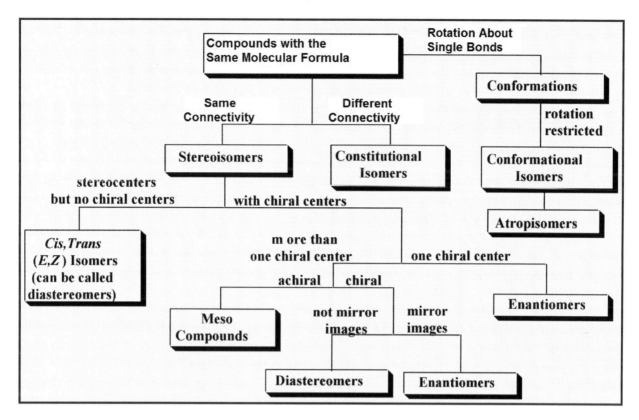

R and S Convention

Each chiral center is designated as either R or S in the IUPAC system.

Steps

- The four different groups attached to the chiral atom are ranked from 1 to 4
    - Where 1 is the highest priority and 4 is the lowest priority
- Chiral center is reoriented (if required) so that the lowest priority group (4) is placed towards the back (into the plane of the paper, away from you)
    - If the other higher priority groups (1, 2, and 3) are in clockwise order the chiral center is designated as **R**
    - If the other higher priority groups (1, 2, and 3) are in counterclockwise order the chiral center is designated as **S**

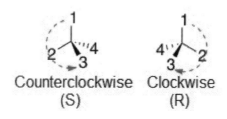

Priority Rules

- Each atom bonded to the chiral center is assigned priority based on atomic number
    - Higher priority is given to atoms with higher atomic number

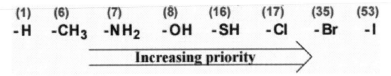

(1)	(6)	(7)	(8)	(16)	(17)	(35)	(53)
-H	-CH$_3$	-NH$_2$	-OH	-SH	-Cl	-Br	-I

Increasing priority →

- Hydrogen is always the lowest priority group if it is attached to the chiral center

- If two groups have the same atom bonded to the chiral center, look at the next set of atoms
    - Priority based on atomic number is assigned at the first point of difference

(1)	(6)	(7)	(8)
-CH$_2$-H	-CH$_2$-CH$_3$	-CH$_2$-NH$_2$	-CH$_2$-OH

Increasing priority →

- Double and triple bonds are treated as if they are a series of single bonds to the same atom

$-CH=CH_2$ is treated as $\underset{}{\overset{C\ \ C}{-CH-CH_2}}$

$-\overset{O}{\underset{}{\overset{\|}{C}}}H$ is treated as $-\overset{O-C}{\underset{H}{\overset{|}{C}-O}}$

$-C\equiv CH$ is treated as $-\overset{C\ \ C}{\underset{C\ \ C}{C-C-H}}$

Fischer Projections

Fischer projections are 2D representations of 3D organic molecules with multiple chiral centers.

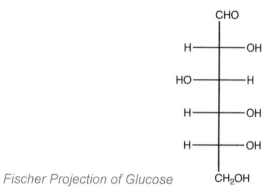

Fischer Projection of Glucose

- Horizontal segments of a Fischer projection represent bonds that are coming towards you

- Vertical segments of a Fischer projections represent bonds that are direct away from you

- Intersections represent a carbon atom

Steps for Converting Line Diagrams to a Fischer Projection

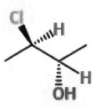

Let's Convert the Following Line Diagram to a Fischer Projection:

- Number the carbons
    - Makes it easier to keep track of the carbons when you start converting

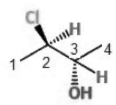

- Identify stereocenters and assign R and S
    - C$_2$ is R
    - C$_3$ is S
- Draw the Fischer projection
    - Chiral carbons are intersection points
    - Draw the groups connected to the chiral carbons
    - Make sure the chiral carbons are R and S in the Fischer projections, as they were when you assigned them on the line angle formula

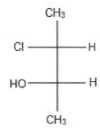

Plane-Polarized Light, Optical Activity, and Racemic Mixtures

- Ordinary light – light that oscillates in all planes perpendicular to its direction of propagation
- Plane-polarized light - light that oscillates only in parallel planes
    - Plane-polarized light is the vector sum of left and right circularly polarized light
        - Circularly polarized light interacts one way with an R chiral center and reacts the opposite way with its enantiomer
- Optically active - refers to a compound that rotates the plane of plane-polarized light
    - Dextrorotatory (+) - refers to a compound that rotates the plane of polarized light to the right
    - Levorotatory (-) - refers to a compound that rotates the plane of polarized light to the left

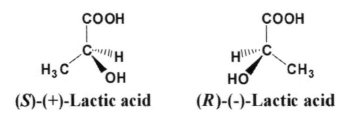

- Racemic mixture – mixture that has equal amounts of left-handed and right-handed enantiomers of a chiral molecule
    - Contains equal amounts of dextrorotatory and levorotatory molecules
        - Thus, it is optically inactive
        - And there is a net zero rotation of plane-polarized light
- Resolution – separation of a racemic mixture into its enantiomers

Chirality in the Biological World

Enzymes

- Each enzyme has an active site where catalysis takes place
    - Active site - a region on the enzyme which binds substrate
- Act only on a specific substance (substrate)

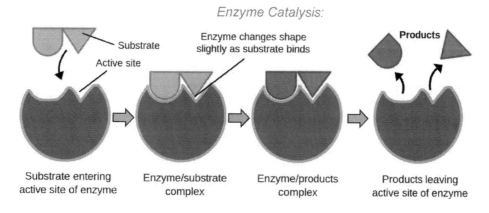

Enzyme Catalysis:

- Left-handed molecule will only fit into a left-handed binding site
- Right-handed molecule will only fit into a right-handed binding site
- Enantiomers have different physiological properties because of the differences in their interactions with other chiral molecules
    - For example, an enzyme may be able to bind with (R)-glyceraldehyde but not with (S)-glyceraldehyde

Amino Acids

- The 20 common amino acids have a central carbon (called the α-carbon)
    - Central carbon is bonded to an amino group (-NH₂) and a carboxylic acid group (-COOH)

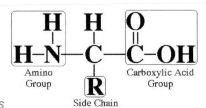

General Structure of Amino Acids

- α-carbon is chiral in 19 out of the 20 common amino acids
    - Glycine does not have a chiral α-carbon
- α-carbon has an S configuration in 18 out of the 19 common amino acids
    - Cysteine has an R configuration
- Zwitterion – molecule or ion that has separate positively and negatively charged groups
    - In amino acids, there is an internal transfer of a hydrogen ion (H⁺) from the carboxylic acid group to the amino group

Amino Acid Zwitterion:

Chapter 4: Acids and Bases

Definitions and Conventions

Acid-base reactions are a type of chemical process typified by the exchange of one or more hydrogen ions (i.e. exchange/transfer of a proton).

- Arrhenius definition
    - Acid – substance that produces H^+ ions in aqueous solution
        - We now know that H^+ reacts immediately with a water molecule to produce a hydronium ion (H_3O^+)
    - Base – substance that produces OH^- ions in aqueous solution
- Bronsted-Lowry definition
    - Acid – proton donor
    - Base – proton acceptor
    - Bronsted-Lowry definition does not require water as a reactant
- Conjugate acids and bases
    - Conjugate base – species that is formed when an acid donates a proton to a base
    - Conjugate acid – species that is formed when a base accepts a proton from an acid
    - Conjugate acid-base pair – pair of molecules or ions that can be interconverted through the transfer of a proton

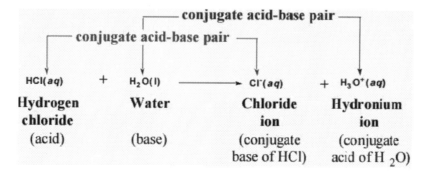

- Curved arrows are used to show the flow of electrons in an acid-base reaction

$$CH_3-C(=O)-\ddot{O}-H + :NH_3 \rightleftharpoons CH_3-C(=O)-\ddot{O}:^- + H-\overset{+}{N}H_3$$

Acetic acid (proton donor)   Ammonia (proton acceptor)   Acetate ion   Ammonium ion

- Resonance and Acids
    - There are many organic molecules that have two or more sites that can act as proton acceptors
        - Preferred site of protonation is the site where the charge is more delocalized

Strengths of Acids and Bases

- Strength of an acid is expressed by an equilibrium constant
    - Equilibrium expression for the dissociation of an uncharged acid (HA)

$$HA + H_2O \rightleftharpoons A^- + H_3O^+$$

$$K_{eq} = \frac{[H_3O^+][A^-]}{[HA][H_2O]}$$

   - K_a, the acid dissociation constant, is given by:

$$K_a = K_{eq}[H_2O] = \frac{[H_3O^+][A^-]}{[HA]}$$

- pK_a and Trends
    - $K_a = 10^{-pKa}$
    - $pK_a = -\log(K_a)$
        - Lower the pK_a, the stronger the acid
            - Higher the pK_a, the weaker the acid
        - Lower the pK_a, the weaker the conjugate base
            - Higher the pK_a, the stronger the conjugate base

- Equilibrium favors the side of the weakest acid and weakest base
  - Equilibrium favors the side with the highest pK_a
    - Thus, pK_a can be used to predict in which direction equilibrium lies

Thermochemistry and Mechanisms

- Reaction mechanism – step-by-step description of how a chemical reaction is occurring
- Thermochemistry – study of energy of the entire system at each step and every instant of a reaction
- Many chemical reactions occur through collisions
  - When collisions occur, the structure of a molecule becomes distorted
    - Higher energy collisions create greater distortions in structure
  - When collisions occur, the kinetic energy of the reactants is converted to potential energy
    - It becomes stored in chemical structures as structural strains
  - During a collision, a transition state ($\ddagger$) is formed
    - In the transition state, there are partially broken and partially formed bonds
- Reaction coordinate diagram – graph showing the energy changes that occur during a chemical reaction

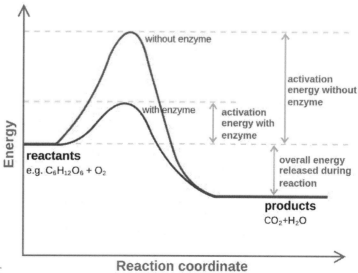

Reaction Coordinate Diagram:

- - Progress of the reaction (time) is indicated on the x-axis
  - Energy is indicated on the y-axis
    - For reaction occurring at a constant pressure, the change in Gibbs free energy, $\Delta G°$ is shown
      - $\Delta G° = -RT \ln(K_{eq})$
      - Free energy of activation – difference in energy between reactants and the transition state

Spontaneous vs. Non-spontaneous Reactions

- Spontaneous if ΔG_{rxn} is negative

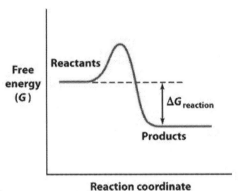

General Spontaneous Reaction Diagram:

- Non-spontaneous if ΔG_{rxn} is positive

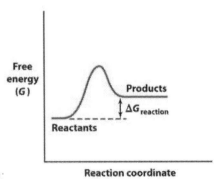

General Non-spontaneous Reaction Diagram:

Molecular Structure and Acidity

- Most important principle in determining the relative acidities of uncharged organic acids is the stability of the anion (A⁻) resulting from the loss of a proton
- The acidity of the acid (HA) is greater if the resulting anion (A⁻) is more stable

- Anions can be stabilized by having the negative charge:
    - On a more electronegative atom
        - Electronegativity – an atom's ability to attract and hold on to electrons, represented by a number
        - Higher electronegativity means the atom holds on to its electrons more strongly which stabilizes the anion
    - On a larger atom
        - There is a greater dispersal of charge
    - Delocalized through resonance
        - Negative charge is disperse over multiple atoms
    - Delocalized by the inductive effect
        - Atoms with high electronegativity are electron withdrawing
            - They pull electron density towards them
            - Electron density on the more electronegative atom is stabilizing as discussed earlier
        - Stabilization by the inductive effect becomes lesser the farther the electronegative atom is from the site of negative charge in the conjugate base
    - In an orbital with more s character
        - Remember from general chemistry that there are different types of hybrid orbitals, or look back at chapter 1 of this guide to review
            - Types: sp, sp^2, sp^3, sp^3d, sp^3d^2
                - sp hybridized orbital has 50% *s* character
                - sp^2 hybridized orbital has 33.33…% *s* character
                - sp^3 hybridized orbital has 25% *s* character, and so on and so forth
                - *s* character (highest to lowest): sp, sp^2, sp^3, sp^3d, sp^3d^2
            - The greater the percentage of *s* character the more stable the anion

CHAPTER 5: ALKENES

Unsaturated Hydrocarbons

Unsaturated hydrocarbons contain at least one or more double or triple bonds.

- Alkene – contains at least one C-C double bond
    - General formula: C_nH_{2n}
- Alkyne – contains at least one C-C triple bond
    - General formula: C_nH_{2n-2}
- Arenes – aromatic hydrocarbons (most commonly based on benzene and its derivatives)

Benzene:

Structure of Alkenes

- Alkene double bonds consists of:
    - One sigma (σ) bond
        - Formed by the overlap of sp^2 hybrid orbitals
    - One pi (π) bond
        - Formed by the overlap of parallel 2p orbitals
    - Two carbon atoms of a double bond and the four atoms bonded to them have bond angles of ~120°

Cis, Trans Isomers (Review)

Cis and trans isomers have the same connectivity but a different arrangement their atoms in space due to the presence of either a ring or a double bond.

- The following are not isomers:

- They are not isomers because there is free-rotation around single bonds
- The following are isomers:

- They are isomers because there is no free-rotation around double bonds
  - Chlorine atoms are locked in their positions
- *Trans* (latin meaning "across") isomers – functional groups are on opposite sides

  trans-1,2-dichloroethene:

- *Cis* (latin meaning "on this side") isomers – functional groups are on the same side

  cis-1,2-dichloroethene:

Index of Hydrogen Deficiency (IHD)

The index of hydrogen deficiency (IHD) is the sum of the number of rings and pi bonds in a molecule.

- IHD for neutral molecules must be an integer
- To determine the IHD, we begin by comparing the number of hydrogens in an unknown compound with the number of hydrogens in a reference hydrocarbon

- Reference hydrocarbon has:
  - Same number of carbons as the unknown compound
  - No rings or pi bonds
  - Molecular formula of C_nH_{2n+2}
- IHD Formula
  - $$IHD = \frac{(H_{reference} - H_{molecule})}{2}$$
  - Things to take into account:
    - For each atom of a Group 7 element (F, Cl, Br, I), you have to add one hydrogen to the reference hydrocarbon
    - No correction has to be made for each atom of a Group 6 element (O, S) to the reference hydrocarbon
    - For each atom of a Group 5 element (N, P), you have to add one hydrogen to the reference hydrocarbon

IUPAC Nomenclature

IUPAC – General (Review)

- Parent chain – longest carbon chain in a molecule
  - The parent name is used to specify the number of carbon atoms in the parent chain

Parent Name	Number of Carbons	Parent Name	Number of Carbons
Meth-	1	Undec-	11
Eth-	2	Dodec-	12
Prop-	3	Tridec-	13
But-	4	Tetradec-	14
Pent-	5	Pentadec-	15
Hex-	6	Hexadec-	16
Hept-	7	Heptadec-	17
Oct-	8	Octadec-	18
Non-	9	Nonadec-	19
Dec-	10	Eicos-	20

- Infix is used to inform about the type of Carbon-Carbon bonds in the parent chain

Infix	Carbon-Carbon Bonds in the Parent Chain
-an-	All single bonds
-en-	One or more double bonds
-yn-	One or more triple bonds

- Suffix is used to inform about the class of compound

Suffix	Class
-e	Hydrocarbon
-ol	Alcohol
-al	Aldehyde
-amine	Amine
-one	Ketone
-oic acid	Carboxylic Acid

- Substituent – group bonded to the parent chain
    - Alkyl group – substituent derived by removal of a hydrogen from an alkane
        - Alkyl groups are symbolized by the capital letter "R"

Common Alkyl Group Substituents:

Alkyl Group	Substituent Name
CH_3-	Methyl
CH_3CH_2-	Ethyl
$CH_3CH_2CH_2-$	Propyl
$-CHCH_3$ $\;\;\;\|$ $\;\;CH_3$	Isopropyl
$CH_3CH_2CH_2CH_2-$	Butyl
$-CH_2CHCH_3$ $\;\;\;\;\;\;\;\;\|$ $\;\;\;\;\;\;CH_3$	Isobutyl
$-CHCH_2CH_3$ $\;\|$ CH_3	Sec-butyl
$\;\;\;CH_3$ $\;\;\;\;\|$ $-CCH_3$ $\;\;\;\;\|$ $\;\;\;CH_3$	Tert-butyl

Naming Alkenes

- Suffix –ene specifies an alkene (e.g., hex<u>ene</u>, pent<u>ene</u>)
- Identify the parent chain (longest carbon chain) and number it (always number sequentially)
    - Numbering must also be done in the direction that gives the carbons of the double bond the lowest number possible
        - Example: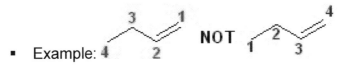
    - Two carbons define the location of the double bond
        - But only the first carbon (lowest number carbon) is used in naming
    - Number of carbons in the parent chain gives you the parent name, then add the suffix –ene, and you add the number of the first carbon with a hyphen in front of the parent name
        - So for the alkene above, there are 4 carbons so the parent name is but- and you would add the suffix –ene to get "butene" and then you add the number of the first carbon with a hyphen in front of the parent name to get "1-butene"
- Each substituent has a name and a number (use a hyphen to connect the name and number)
    - Number of the substituent is determined by which carbon it is on
        - Example: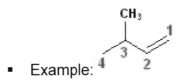
        - Methyl group (CH$_3$-) is on C$_3$ so the substituent would be named 3-methyl
        - You combine the name and number of the substituent with the parent name so the name of the compound would be "3-methyl-1-butene"

- If there are two or more of the same substituent, add a comma to separate the substituent numbers and add a prefix to indicate how many of the substituents you have
    - Two of the same substituent (di-)
    - Three of the same substituent (tri-), and so on and so forth
- If there are two or more different substituents:
    - List them in alphabetical order
- Unlike alkanes where numbering the parent chain must be done so that substituents get the smallest possible numbers, alkenes must always be numbered so the carbons in the double bonds get the smallest possible numbers
- For alkenes containing two or more double bonds the infix "-en" is changed
    - –adien- for two double bonds

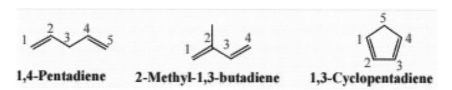

1,4-Pentadiene   2-Methyl-1,3-butadiene   1,3-Cyclopentadiene

- –atrien- for three double bonds, and so on and so forth

Assigning *Cis* and *Trans*

- Many alkenes exist in two isomeric forms, yet have the same connectivity of atoms (constitutional isomers)
    - These isomers differ in the orientation of groups and are geometric isomers
        - Example: 2-butene

cis-2-butene   *trans*-2-butene

- To determine *cis* or *trans*, use the double bond as a reference plane
  - If the two alkyl groups are on the same side of the double bond then it is *cis*
  - If the two alkyl groups are on the opposite side of the double bond then it is *trans*

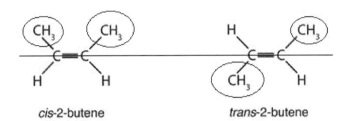

cis-2-butene   trans-2-butene

E and Z Nomenclature of Alkenes

The E and Z nomenclature is more reliable and more suited than the *cis* and *trans* system to be applied to tri- or tetra-substituted alkenes, especially when the substituents are not alkyl groups.

- This system uses priority rules that were discussed in Chapter 3 of this guide
  - Each atom bonded to the chiral center is assigned priority based on atomic number
    - Higher priority is given to atoms with higher atomic number

$$\underset{\text{Increasing priority} \longrightarrow}{\overset{(1)\quad(6)\quad(7)\quad(8)\quad(16)\quad(17)\quad(35)\quad(53)}{-H\quad -CH_3\quad -NH_2\quad -OH\quad -SH\quad -Cl\quad -Br\quad -I}}$$

   - Hydrogen is always the lowest priority group if it is attached to the chiral center
  - If two groups have the same atom bonded to the chiral center, look at the next set of atoms
    - Priority based on atomic number is assigned at the first point of difference

$$\underset{\text{Increasing priority} \longrightarrow}{\overset{(1)\qquad\qquad(6)\qquad\qquad(7)\qquad\qquad(8)}{-CH_2-H\quad -CH_2-CH_3\quad -CH_2-NH_2\quad -CH_2-OH}}$$

- Double and triple bonds are treated as if they are a series of single bonds to the same atom

$$-CH=CH_2 \xrightarrow{\text{is treated as}} \begin{array}{c} C\ \ C \\ |\ \ \ | \\ -CH-CH_2 \end{array}$$

$$\underset{-CH}{\overset{O}{\|}} \xrightarrow{\text{is treated as}} \begin{array}{c} O-C \\ | \\ -C-O \\ | \\ H \end{array}$$

$$-C\equiv CH \xrightarrow{\text{is treated as}} \begin{array}{c} C\ \ C \\ |\ \ \ | \\ -C-C-H \\ |\ \ \ | \\ C\ \ C \end{array}$$

- If the higher priority groups are on the same side, the configuration is Z
- If the higher priority groups are on the opposite side, the configuration is E

High Priority H₃C   CH₃ High Priority | High Priority H₃C   H Low Priority
            C=C | C=C
Low Priority H   H Low Priority | Low Priority H   CH₃ High Priority
            (Z)-2-butene | (E)-2-butene

CHAPTER 6: REACTIONS OF ALKENES

Gibbs Free Energy

Gibbs free energy can be used to determine the direction of the chemical reaction under given conditions.

- $\Delta G = \Delta H - T\Delta S$ or $\Delta G = G_{products} - G_{reactants}$
  - G = Gibbs free energy (J/mol)
  - H = enthalpy (J/mol) - total energy content of a system
  - S = entropy (J/K*mol) - measure of disorder or randomness (how energy is dispersed)
  - T = Temperature (K)
    - As T increases so does S
- $+\Delta G$ means energy must be put into the system
  - Indicates that a process is **nonspontaneous or endergonic**
  - Indicates that the position of the equilibrium for a reaction favors the products
- $-\Delta G$ means energy is released by the system
  - Indicates that a process is **spontaneous or exergonic**
  - Indicates that the position of the equilibrium for a reaction favors the reactants
- $\Delta G = 0$ indicates that the system is at equilibrium

Important: ΔG only indicates if a process occurs spontaneously or not, but does **not** indicate anything about how fast a process occurs.

Energy Diagrams

Energy diagrams are graphs showing the changes in energy that occur during a chemical reaction.

- Transition state (‡)
  - Unstable species with high energy formed during a reaction

- Activation energy ($\Delta G^{\ddagger}$) – difference in Gibbs free energy between reactants and a transition state
    - If $\Delta G^{\ddagger}$ is large; the reaction occurs slowly
        - Few collisions occur with sufficient energy to reach the transition state
    - If $\Delta G^{\ddagger}$ is small; the reaction occurs quickly
        - Many collisions occur with sufficient energy to reach the transition state
- One-step reaction

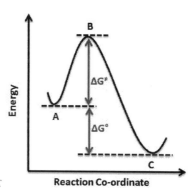

One-step Reaction Energy Diagram:

    - One transition state
- Two-step reaction

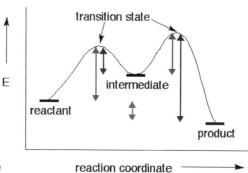

Two-step Reaction Energy Diagram

    - Two transition states
    - One intermediate

Reaction Mechanisms

A reaction mechanism is step-by-step description of how a chemical reaction is occurring. It shows the following:

- Which bonds are broken
- Which bonds are formed
- Order and relative rated of the various bond-breaking and bond-forming steps
- If the reaction occurs in a solution, the role of the solvent
- If there is a catalyst, the role of a catalyst
- Position of all atoms
- Energy of the entire system during the reaction

Electron Pushing

Electron pushing (aka arrow pushing) is used to depict the flow of electrons during a chemical reaction.

- Arrows are used to indicate movement of electrons

curved, two-barbed arrow:
two electron movement

curved, single-barbed ('fish-hook') arrow:
single electron movement

- Arrows are never used to indicate the movements of atoms
    - Make sure an arrow starts from electrons or bonds, never from an atom
- Arrows always start at an electron source
    - Electron source – most commonly, either a π bond or a lone pair of electrons
- Arrows always end at an electron sink
    - Electron sink – an atom or ion that can accept a new bond or a lone pair of electrons
        - Keep in mind, bonds may have to be broken to avoid overfilling the valence shell of an atom serving as an electron sink

Patterns of e- Movement

- Electrons are moved for these typical reasons:
    - Redistribution of π bonds and/or lone pairs
    - Forming a new σ bond from a lone pair or a π bond
    - Breaking a σ bond to form a π bond or a new lone pair

Common Mechanisms

These are four of the most common mechanism elements to consider when predicting the individuals steps of a chemical reaction:

- Attack of the nucleophile
    - Make a new bond between a nucleophile (electron source) and an electrophile (electron sink)
- Departure of the leaving group
    - Break a bond so that a relatively stable ion or molecule is formed
- Add a proton
    - Used when there is no suitable nucleophile-electrophile reaction
    - But a molecule has a strongly basic functional group
    - Or a strong acid is present
- Remove a proton
    - Used when there is no suitable nucleophile-electrophile reaction
    - But a molecule has a strongly acidic proton
    - Or a strong base is present

Carbocations

A carbocation is a carbon atom with only six electrons in its valence shell and a positive charge.

- Are electrophiles ("electron loving")
- Are Lewis acids

- Are classified depending on the number of carbons that are attached to them
    - 1° carbocation – carbocation attached to one carbon
    - 2° carbocation – carbocation attached to two carbons
    - 3° carbocation – carbocation attached to three carbons
- Use sp² hybrid orbitals to form sigma bonds from carbon to the three attached groups
    - Unhybridized 2p orbital lies perpendicular to the sigma bond framework and contains no electrons

Carbocation Stability

- Carbocation stability: 3° > 2° > 1°
    - 3° is more stable than 2°
    - 2° is more stable than 1°

most stable ⟹ least stable

CH₃	CH₃	CH₃	H
H₃C–C⁺–CH₃	H–C⁺–CH₃	H–C⁺–H	H–C⁺–H
tertiary	secondary	primary	methyl

- Alkyl groups bonded to a positively charged carbon help delocalize the positive charge of the cation (this is their "electron-releasing" ability)

Alkyl Group Stability Effect (Most Stable Carbocation to the Least Stable):

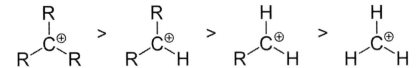

- Electron-releasing ability of alkyl groups is due to:
    - Inductive effect
        - Positively charged carbon polarizes electrons of adjacent sigma bonds toward it
        - Positive charge on the carbocation becomes delocalized over nearby atoms

- Stability of the cation increases with the amount of volume over which the charge is delocalized
  - Hyperconjugation
    - Partial overlap of the sigma-bonding orbital of an adjacent C-H or C-C bond with the vacant 2p orbital of the carbocation that gives an extended molecular orbital that increases the stability of the system

Reaction: Addition of H-X to an Alkene – Electrophilic Addition Reaction

- X is a stand-in for a Group 7 element (e.g., Cl, Br)
- Addition is regioselective
  - Regioselective reaction – addition or substitution reaction in which one product is preferred above all other
  - *Important*: Markovnikov's Rule – the addition of HX or H_2O to an alkene results in hydrogen adding to the carbon of the double bonds that has the greater number of hydrogens
- Mechanism steps:
  - **Step 1:** Add a proton
    - Proton transfer from HX to the alkene gives a carbocation intermediate

- There are two possibilities in terms of which carbon the positive charge can end up on
  - Probability #2 is more likely because a 2° carbocation is more stable than a 1° carbocation
    - **Step 2:** Attack of the nucleophile
      - Reaction of the cation with the anion completes the reaction

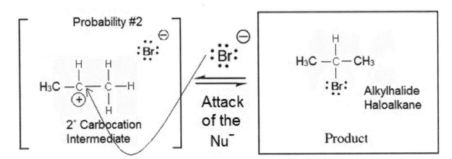

- Reaction Summary: π bond is protonated in the first step to form a carbocation. The most stable carbocation is formed. The anion attacks the carbocation to give an alkylhalide (aka haloalkane).

$$\text{Alkene} \xrightarrow{\text{HX}} \text{Alkylhalide (Haloalkane)}$$

  - Reaction starts with an alkene
    - Transformed to an alkylhalide (aka haloalkane) by the end of the reaction
  - Regioselectivity: Markovnikov
    - Hydrogen adds preferentially to the less substituted carbon of the double bond (the carbon atom bearing the greater number of hydrogens)
  - Stereoselectivity: Not a stereoselective reaction
  - Carbocation rearrangement is possible (carbocation rearrangement is discussed later in this chapter)

Reaction: Acid Catalyzed Hydration of an Alkene

- Addition of water (H_2O) is called hydration

- Mechanism steps:
    - **Step 1:** Add a proton

    - **Step 2:** Attack of the nucleophile

    - **Step 3:** Remove a proton

- Reaction Summary: π bond adds a proton to give a stable carbocation. Carbocation reacts with the nucleophile, water, to give an oxonium ion intermediate. The oxonium ion intermediate loses a proton to form an alcohol.

$$\text{Alkene} \xrightarrow[\text{cat } H_2SO_4]{H_2O} \text{Alcohol}$$

  - Reaction starts with an alkene
    - Transformed to an alcohol by the end of the reaction
  - Reaction is catalyzed by H_2SO_4
    - H_2SO_4 provides the hydrogen needed to convert water (H_2O) to hydronium (H_3O^+)
      - Note that in the first step the proton is being added from hydronium
  - Regioselectivity: Markovnikov
    - Hydrogen adds preferentially to the less substituted carbon of the double bond (the carbon atom bearing the greater number of hydrogens)
  - Stereoselectivity: Not a stereoselective reaction
  - Carbocation rearrangement is possible

Carbocation Rearrangement

In electrophilic addition to alkenes, there is the possibility for rearrangement if a carbocation is involved.

- Rearrangement – change in connectivity of the atoms in a product compared with the connectivity of the same atoms in the starting material
- Driving force in a carbocation rearrangement is that a less stable carbocation is rearranged to a more stable carbocation

Reaction: Carbocation Rearrangement – Addition of H-X to an Alkene – 1,2 Methyl Shift

- A methyl group can move with its electrons from one atom to another adjacent electron deficient carbocation
  - Driving force for this is that a less stable carbocation is converted to a more stable carbocation

- Mechanism steps
    - **Step1:** Add a proton

    - Step 2: 1,2 Methyl Shift – Carbocation Rearrangement

    - Step 3: Attack of the nucleophile

Reaction: Carbocation Rearrangement - 1,2 Hydrogen Shift - Acid Catalyzed Hydration of an Alkene

- Carbocation can rearrange if moving an adjacent hydrogen atom leads to a more stable carbocation

- Mechanism steps
    - **Step 1:** Add a proton

    - **Step 2:** 1,2 Hydrogen Shift – Carbocation Rearrangement

    - **Step 3:** Attack of the nucleophile

- Step 4: Remove a proton

Reaction: Acid Catalyzed Addition of an Alcohol to an Alkene

- Mechanism steps
    - **Step 1:** Add a proton

    - **Step 2:** Attack of the nucleophile

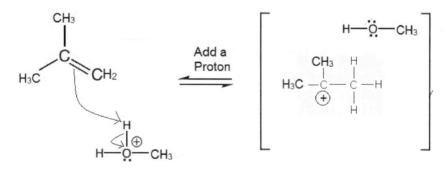

- **Step 3:** Remove a proton

- Reaction Summary:

  Alkene $\xrightarrow[\text{cat } H_2SO_4]{ROH}$ Ether

    - Reaction starts with an alkene
        - Transformed to an ether by the end of the reaction
    - Reaction is catalyzed by H_2SO_4
        - H_2SO_4 provides the hydrogen needed to convert CH_2OH to CH_3OH^+
            - Note that in the first step the proton is being added from CH_3OH^+
    - Regioselectivity: Markovnikov
        - Hydrogen adds preferentially to the less substituted carbon of the double bond (the carbon atom bearing the greater number of hydrogens)
    - Stereoselectivity: Not a stereoselective reaction
    - Carbocation rearrangement is possible

Reaction: Addition of X_2 to Alkene

- Typically this is an addition of Cl_2, Br_2, or I_2
- Carried out with either pure reagents or in an inert solvent such as CH_2Cl_2

- Mechanism steps
    - **Step 1:** Electrophilic addition
        - Electrophilic addition is a reaction between an electrophile and a nucleophile, adding to double or triple bonds
        - One of the pi bonds is removed and creates two new sigma bonds
        - A bridged bromonium ion intermediate is formed

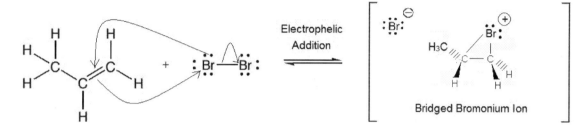

    - **Step 2:** Attack of the nucleophile
        - The halide ion (the nucleophile) attacks from the **opposite** side of the bromonium ion and opens the three-membered ring to create the product

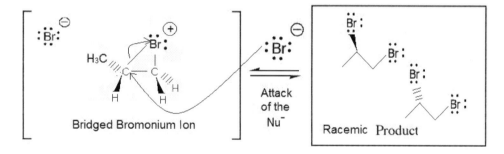

- Reaction Summary:

$$\text{Alkene} \xrightarrow[\text{CH}_2\text{Cl}_2]{X_2} \text{Vicinal Dihalide}$$

    - Reaction starts with an alkene
        - Transformed to a vicinal dihalide by the end of the reaction
            - Vicinal – refers to any two functional groups bonded to two adjacent carbon atoms
    - CH_2Cl_2 is an inert solvent
        - May not be used in some cases

- Regioselectivity: Non-Markovnikov
  - No addition of hydrogen
- Stereoselectivity: Anti-product
  - Halide ion always attacks from the **opposite** side of the bromonium ion because it is sterically hindered from attacking from the same side
- No carbocation rearrangement in this reaction

Reaction: Oxymercuration – Reduction of Alkenes

- First step of the reaction requires Hg(OAc)$_2$, mercury (II) acetate

$$H_3C-\overset{O}{\overset{\|}{C}}-O-Hg-O-\overset{O}{\overset{\|}{C}}-CH_3$$

Hg(OAc)$_2$

Hg(OAc)$_2$ with OAc Identified: OAc   OAc

- Mechanism steps
  - The following occurs before the first step in the mechanism that you are likely to be responsible for:

  - **Step 1:** Attack of the nucleophile
    - Creates a mercurinium ion intermediate

- **Step 2:** Attack of the nucleophile

$$\left[\begin{array}{c} H_3C \overset{+}{\underset{H}{\diagdown}} C - C \overset{Hg-OAc}{\underset{H}{\diagup}} \end{array} \right]$$

H—Ö—H ⇌ Attack of the Nu⁻

$$\left[\begin{array}{c} CH_3 \quad Hg-OAc \\ H-C-C-H \\ \overset{+}{O} \quad H \\ H \quad H \end{array} \right]$$

- **Step 3:** Remove a proton

$$\left[\begin{array}{c} CH_3 \quad Hg-OAc \\ H-C-C-H \\ \overset{+}{O} \quad H \\ H \quad H \end{array} \right] \quad H-\ddot{O}-H \rightleftharpoons \quad \left[\begin{array}{c} CH_3 \quad Hg-OAc \\ H-C-C-H \\ H\ddot{O}: \quad H \end{array} \right]$$

Remove a Proton

- **Step 4:** Chemist opens the flask and adds a reagent (in this reaction, NaBH₄)
    - Result of this step: NaBH₄ is a reducing agent, the bond with mercury is replaced with a hydrogen

$$\left[\begin{array}{c} CH_3 \quad Hg-OAc \\ H-C-C-H \\ H\ddot{O}: \quad H \end{array} \right]$$

NaBH₄ ⇅ (Chemist opens flask and adds new reagent)

$$\begin{array}{c} CH_3 \quad H \\ H-C-C-H \\ H\ddot{O}: \quad H \end{array}$$

Products

- Reaction Summary: π bond attacks the mercury acetate cation and gives a bridged mercurinium ion intermediate. H₂O (a nucleophile) attacks the more substituted carbon from the opposite side. Removal of the proton and reduction with NaBH₄ gives the final product, an alcohol.

$$\text{Alkene} \xrightarrow[\text{2) NaBH}_4]{\text{1) Hg(OAc)}_2} \text{Alcohol}$$

  - Reaction starts with an alkene
    - Transformed to an alcohol by the end of the reaction
  - Regioselectivity: Markovnikov
    - Hydrogen adds preferentially to the less substituted carbon of the double bond (the carbon atom bearing the greater number of hydrogens)
    - Also note that the OH group of the final product is added to the other carbon (the more substituted Carbon)
  - Stereoselectivity: Overall, not a stereoselective reaction
  - No carbocation rearrangement in this reaction

Reaction: Hydroboration Oxidation

- First step of the reaction requires BH₃, borane
- Mechanism steps
  - **Step 1:** Simultaneous bond formation
    - Regioselective and stereoselective addition of B and H to the carbon-carbon double bond

    (R = H or alkyl group depending on how far reaction has progressed)

    transition state

    - In the figure above, you are looking at the transition state and not the intermediate

- Dashed bond lines (not to be confused with dashed wedges) are showing bonds that are forming and bonds that are breaking
  - Boron adds to the carbon with less steric hindrance

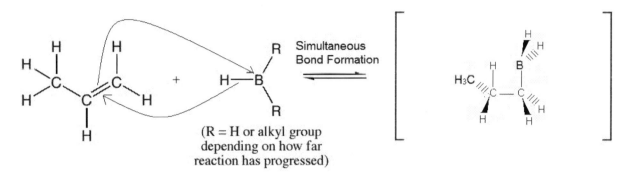

(R = H or alkyl group depending on how far reaction has progressed)

  - In the figure above, you are looking at the intermediate
- **Step 2:** Chemist opens the flask and adds H_2O_2/OH^-
  - Result of this step: bond with boron is replaced with a hydroxyl group (-OH)

- Reaction Summary: there is simultaneous addition of the boron and the hydrogen on the same side of the double bond. H_2O_2 (hydrogen peroxide) replaces the boron with an OH group to produce an alcohol.

$$\text{Alkene} \xrightarrow[\text{2) } H_2O_2/OH^-]{\text{1) } BH_3} \text{Alcohol}$$

- Reaction starts with an alkene
  - Transformed to an alcohol by the end of the reaction
- Regioselectivity: Non-Markovnikov
- Stereoselectivity: Syn-product
  - Hydroxyl group (-OH) and the hydrogen add to the same side
- No carbocation rearrangement in this reaction

Reaction: Osmium Tetroxide – Oxidation of Alkenes to Glycol (Diol)

- Osmium tetroxide (OsO_4) oxidizes an alkene to a glycol
    - Glycol (aka diol) – compound with OH groups on adjacent carbons
    - OsO_4 is expensive and highly toxic
        - So it is used in catalytic amounts with another oxidizing agent to reoxidize its reduced forms (recycles OsO_4)
- Mechanism steps
    - **Step 1:** Electrophilic addition

    - **Step 2:** Chemist opens the flask and adds the reducing agent ($NaHSO_3/H_2O$)
        - Result of this step: cyclic osmate intermediate is reduced by the reducing agent which cleaves the osmium-oxygen bond to give a vicinal diol
            - Vicinal – refers to any two functional groups bonded to two adjacent carbon atoms

- Reaction Summary: alkene and OsO₄ form a cyclic osmate intermediate. Cyclic osmate intermediate is reduced by the reducing agent which cleaves the osmium-oxygen bond to give a vicinal diol
    - Vicinal – refers to any two functional groups bonded to two adjacent carbon atoms

$$\text{Alkene} \xrightarrow[\text{2) NaHSO}_3/\text{H}_2\text{O}]{\text{1) OsO}_4} \text{Vicinal Diol}$$

    - Reaction starts with an alkene
        - Transformed to a vicinal diol by the end of the reaction
    - Regioselectivity: Non-Markovnikov
    - Stereoselectivity: Syn-product
        - Both hydroxyl groups (-OH) add to the same face
    - No carbocation rearrangement in this reaction

Reaction: Ozonolysis

- First step of the reaction requires ozone (O₃)
- Mechanism steps (this is a partial mechanism)
    - **Step 1:** Electrophilic addition
        - Creates a malozonide complex

- **Step 2:** Two-step rearrangement
    - Carbon-carbon bond is broken
    - Malozonide complex rearranges in two steps to an ozonide complex

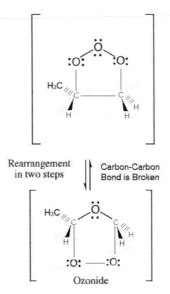

- **Step 3:** Chemist opens flask and adds the reducing agent
    - Reducing agent of this reaction is (CH$_3$)$_2$S, dimethyl sulfide
    - Result of this step: ozonide complex is reduced by dimethyl sulfide into an aldehyde

- Reaction Summary: oxidation of alkenes

$$\text{Alkene} \xrightarrow[\text{2) }(CH_3)_2S]{\text{1) }O_3} \text{Aldehyde/Ketone}$$

  - Reaction starts with an alkene
    - Transformed to aldehyde/ketone by the end of the reaction
    - The original C-C double bond is completely gone by the end of the reaction
    - If this reaction occurs with a double bond inside a ring structure, the loss of the C-C double bond opens up the ring structure

  Ozonolysis with a Ring-Structure:

  - Regioselectivity: Non-Markovnikov
  - Stereoselectivity: Not a stereoselective reaction
  - No carbocation rearrangement in this reaction

Reaction: Hydrogenation of Alkenes

- This reaction is also known as "reduction of alkenes" and "catalytic hydrogenation"
- In the presence of a transition metal catalyst, most alkenes react with H_2 to produce alkanes
  - Common transition metal catalysts are: Pt, Pd, Ru, and Ni
- Reduction of an alkene to an alkane is exothermic
  - Net conversion of one pi bond to two sigma bonds
  - A *trans* alkene is more stable than a *cis* alkene

- Mechanism

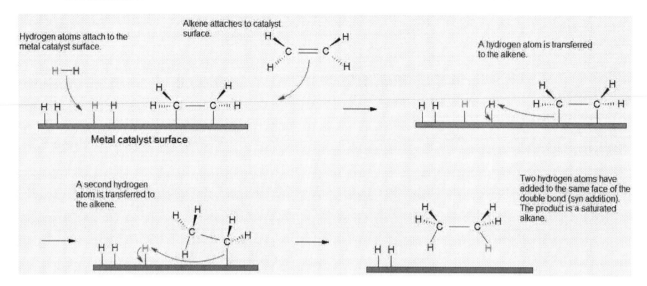

- Reaction summary: H_2 gas adheres on the transition metal surface and makes sigma H-metal bonds. The alkene adheres onto the metal and makes sigma C-metal bond. Addition of hydrogen atoms takes place from the same side.

$$\text{Alkene} \xrightarrow[\text{Pd/Pt}]{H_2} \text{Alkane}$$

  o Reaction starts with an alkene
    - Transformed to an alkane by the end of the reaction
  o Regioselectivity: Non-Markovnikov
  o Stereoselectivity: Syn-stereoselectivity
    - Hydrogens add to the same face
  o No carbocation rearrangement in this reaction

CHAPTER 7: ALKYNES AND REACTIONS OF ALKYNES

Alkynes and their Properties

Alkynes are unsaturated hydrocarbons containing at least one triple bond.

- Nonpolar
- Dissolve in organic solvents
    - Slight solubility in polar solvents
    - Insoluble in water
- Slightly higher boiling point than alkanes and alkenes
- Acidity of terminal alkynes is higher than alkanes and alkenes

IUPAC Nomenclature

Naming Alkynes

- Suffix –yne specifies an alkyne (e.g., hex<u>yne</u>, pent<u>yne</u>)
- Identify the parent chain (longest carbon chain) and number it (always number sequentially)
    - Numbering must also be done in the direction that gives the carbons of the triple bond the lowest number possible
        - Example: $H-\underset{1}{C}\equiv\underset{2}{C}-\underset{3}{CH_2}-\underset{4}{CH_3}$
    - Two carbons define the location of the triple bond
        - But only the first carbon (lowest number carbon) is used in naming
    - Number of carbons in the parent chain gives you the parent name, then add the suffix –yne, and you add the number of the first carbon with a hyphen in front of the parent name
        - So for the alkyne above, there are 4 carbons so the parent name is but- and you would add the suffix –yne to get "butyne" and then you add the number of the first carbon with a hyphen in front of the prefix to get "1-butyne"

- Each substituent has a name and a number (use a hyphen to connect the name and number)
    - Number of the substituent is determined by which carbon it is on
        - Example: [structure of 3-methyl-1-butyne with H₃C, H₃C groups and ≡CH]
        - Methyl group (CH$_3$-) is on C$_3$ so the substituent would be named 3-methyl
        - You combine the name and number of the substituent with the parent name so the name of the compound would be "3-methyl-1-butyne"
    - If there are two or more of the same substituent, add a comma to separate the substituent numbers and add a prefix to indicate how many of the substituents you have
        - Two of the same substituent (di-)
        - Three of the same substituent (tri-), and so on and so forth
    - If there are two or more different substituents:
        - List them in alphabetical order
    - Unlike alkanes where numbering the parent chain must be done so that substituents get the smallest possible numbers, alkynes must always be numbered so the carbons in the triple bonds get the smallest possible numbers
- There may be cases where there is both a double bond and triple bond present in the compound that you are trying to name
    - Start the numbering of the compound from the end closest to the first multiple bond
    - If both the double bond and triple bond are located similar distances from either end of the compound, the double bond gets higher priority (i.e. start numbering from the end closest to the double bond)

Preparation of Alkynes

Creating an Alkyne Anion:

$$H-C\equiv C-H + Na^+ \; :\ddot{N}H_2^- \rightleftharpoons H-C\equiv C:^-$$

Sodium Amide → Acetylide Anion

- Acetylide is a carbanion that is also an alkyne anion
- Alkyne anions are good nucleophiles and strong bases
    - They participate in nucleophilic substitution reactions with alkyl halides to form C-C bonds to alkyl groups
        - Since alkyne anions are also strong bases, alkylation is practical only with methyl and primary (1°) halides
        - With 2° and 3° halides, the major reactions are elimination reactions

Elimination Reaction:

HC≡C⁻ Na⁺ + Bromocyclohexane → HC≡CH + Cyclohexene + Na⁺Br⁻
(Sodium acetylide)   (Acetylene)

- Alkylation of alkyne anions is a convenient method for the synthesis of terminal alkynes

Synthesis of a Terminal Alkyne:

HC≡C:⁻ Na⁺ + 1-Bromobutane → 1-Hexyne + Na⁺Br⁻
(Sodium acetylide)

- A terminal alkyne can be converted to an internal alkyne if alkylation is repeated

Conversion of a Terminal Alkyne to an Internal Alkyne:

$CH_3CH_2C\equiv C^-\; Na^+$ + $CH_3CH_2\text{-}Br$ → $CH_3CH_2C\equiv CCH_2CH_3$ + Na^+Br^-

Sodium butynide   Bromoethane   3-Hexyne

Preparation of Alkynes from Alkenes

- Reaction of a vicinal dibromoalkane with two moles of a base results in two successive dehyrohalogenation reactions (removal of H and X from adjacent carbons) and formation of an alkyne
    - Most commonly used base is sodium amide, $NaNH_2$

    - **Step 1:** $CH_3CH=CHCH_3$ (2-Butene) $+ Br_2 \xrightarrow{CH_2Cl_2}$ $CH_3\overset{Br}{C}H\text{-}\overset{Br}{C}HCH_3$ $+ 2NaNH_2$ (Sodium amide)

    - **Step 2:** $CH_3\overset{Br}{C}H\text{-}\overset{Br}{C}HCH_3 + 2NaNH_2$ (Sodium amide) $\xrightarrow{NH_3(l),\ -33°C}$ $CH_3C\equiv CCH_3$ (2-Butyne) $+ 2NaBr + 2NH_3$

- To synthesize a terminal alkyne from a terminal alkene, 3 moles of base are required

    - **Step 1:** $CH_3(CH_2)_3CH=CH_2$ (1-Hexene) $\xrightarrow{Br_2}$ $CH_3(CH_2)_3\overset{Br}{C}H\text{-}\overset{Br}{C}H_2$ (1,2-Dibromohexane)

    - **Step 2:** $CH_3(CH_2)_3\overset{Br}{C}H\text{-}\overset{Br}{C}H_2$ (1,2-Dibromohexane) $\xrightarrow{3NaNH_2,\ 2HBr}$ $CH_3(CH_2)_3C\equiv C^-Na^+$ (Sodium salt of 1-hexyne)

    - **Step 3:** $CH_3(CH_2)_3C\equiv C^-Na^+$ (Sodium salt of 1-hexyne) $\xrightarrow{H_2O}$ $CH_3(CH_2)_3C\equiv CH$ (1-Hexyne)

Reaction: Addition of HX to an Alkyne

- X is a stand-in for a Group 7 element (e.g., Cl, Br)
- Mechanism steps (using HBr as an example)
    - **Step 1:** Add a proton
        - This step requires 1 mole of HBr

        $H_3C-C\equiv C-H \xrightarrow{\text{Add a Proton, 1 Mole H-Br}} \left[\begin{array}{c} H_3C \\ \end{array} C=C \begin{array}{c} H \\ H \end{array} \right]^+ \quad :Br:^-$

        - Product of this step is a unstable vinylic carbocation

- There are two possibilities in terms of which carbon the positive charge can end up on
  - A 2° vinylic carbocation is preferred over a 1° vinylic carbocation
    - 2° vinylic carbocation is connected to two carbons
    - 1° vinylic carbocation is connected to one carbon
- **Step 2:** Attack of the nucleophile

$$\left[\begin{array}{c} H_3C \\ C = C \\ \oplus H H \end{array} \right]$$

$:\ddot{Br}:^{\ominus}$ ⇅ Attack of the Nu⁻

$$\left[\begin{array}{c} H_3C \\ C = C \\ :\ddot{Br}: H H \end{array} \right]$$

- **Step 3:** Add a proton
  - This step requires another mole of HBr

$$\left[\begin{array}{c} H_3C \\ C = C \\ :\ddot{Br}: H H \end{array} \right] \xrightarrow[\text{1 Mole } H-\ddot{Br}:]{\text{Add a Proton}} \left[H_3C - \overset{\oplus}{\underset{:\ddot{Br}:}{C}} - \overset{H}{\underset{H}{C}} - H \longleftrightarrow H_3C - \overset{}{\underset{:\ddot{Br}:\oplus}{C}} = \overset{H}{\underset{H}{C}} - H \right]$$

  - Note that the positive charge ends up on the more substituted carbon
    - 2° carbocation is more stable than a 1° carbocation
  - Positive charge is stabilized through resonance

- **Step 4:** Attack of the nucleophile

    - Final product with 2 moles of HBr is a geminal dihalide

- Reaction Summary: addition of a proton to the alkyne forms an unstable vinylic carbocation. Addition of another mole of HX forms a 2° carbocation with the positive charge on the carbon bearing the halogen which is stabilized by resonance. The final product with 2 moles of HX is a germinal dihalide.

$$\text{Alkyne} \xrightarrow[\text{HX}]{\text{2 Moles}} \text{Geminal Dihalide}$$

    - Reaction starts with an alkyne
        - Transformed to a germinal dihalide by the end of the reaction
    - Markovnikov regioselectivity can be used to determine which carbon the hydrogens are adding during the reaction
        - Hydrogen adds preferentially to the less substituted carbon (carbon atom bearing the greater number of hydrogens)

Reaction: Addition of X₂ to an Alkyne

- X is a stand-in for a Group 7 element (e.g., Cl, Br)
- Mechanism steps (example using Br₂)
  - **Step 1:** Reaction with 1 mole of Br₂

    - The halide ion (the nucleophile) attacks from the **opposite** side of the bromonium ion and opens the three-membered ring

    - Product of the reaction between an alkyne and 1 mole of Br₂ produces a dibromoalkene

  - **Step 2:** Reaction with another 1 mole of Br₂

    - Product of the reaction between the dibromoalkane and 1 mole of Br₂ produces a tetrahaloalkane

- Reaction Summary

  Alkyne $\xrightarrow{\text{1 mole } X_2}$ Dihaloalkane

  Alkyne $\xrightarrow{\text{2 mole } X_2}$ Tetrahaloalakane

  - Stereoselectivity: Anti-product
  - No carbocation rearrangement in this reaction

Reduction of Alkynes

Reduction of Alkynes with H_2 in the Presence of a Metal Catalyst

- Reaction of alkynes with H_2 in the presence of a metal catalyst (commonly, Pd or Pt) converts the alkyne to an alkane

Reduction of Alkyne with H_2 and a Metal Catalyst:

$$H_3C-C\equiv C-CH_3 \xrightarrow{H_2, \text{ Pd or Pt}} \begin{array}{c} CH_3 \ CH_3 \\ | \ | \\ H-C-C-H \\ | \ | \\ H \ H \end{array}$$

General Reduction of Alkyne with H_2 and a Metal Catalyst:

$$\text{Alkyne} \xrightarrow{H_2, \text{ Pd or Pt}} \text{Alkane}$$

- This is not a selective reduction
  - It results in the **complete** reduction of the triple bond

Reduction of Alkynes by Lindlar's Catalyst

- Lindlar's catalyst is what is referred to as a "poisoned" metal catalyst
  - It lacks the normal activity that we associate with palladium (Pd) catalysts for reducing double bonds (i.e. Lindlar's catalyst can't reduce a double bond)
    - Useful for when you want to create an alkene from an alkyne

Reduction of Alkyne with H_2 and Lindlar's Catalyst:

$$H_3C-C\equiv C-CH_3 \xrightarrow{H_2, \text{ Lindlar's Catalyst}} \begin{array}{c} H_3C \\ \end{array}\!\!\!\!C=C\!\!\!\!\begin{array}{c} CH_3 \\ H \end{array}$$

General Reduction of Alkyne with H_2 and Lindlar's Catalyst:

$$\text{Alkyne} \xrightarrow{H_2, \text{ Lindlar's Catalyst}} \text{Alkane}$$

- This reduction has *syn* stereoselectivity
  - Hydrogens add to the same side

Reduction of Alkynes by Sodium in Liquid Ammonia

- Alkynes can be reduced to *trans* alkenes using Na in $NH_{3(l)}$
- Mechanism steps
    - **Step 1:** One-electron reduction

        $$H_3C-C\equiv C-CH_3 \ + \ \cdot Na \ \xrightleftharpoons{\text{One-Electron Reduction}} \ \left[H_3C-\overset{\cdot}{C}=\overset{\ominus}{\underset{..}{C}}-CH_3 \right]^{\oplus Na}$$

        - Note that single-barbed arrows are being used
            - They indicate the movement of only one electron
        - Sodium transfers one electron to the alkyne which produces a radical anion

    - **Step 2:** Add a proton

        The radical anion is protonated by NH_3 to give a vinyl radical:

        $$\left[H_3C-\overset{\cdot}{C}=\overset{\ominus}{\underset{..}{C}}-CH_3 \right]^{\oplus Na} + H-\overset{..}{\underset{H}{N}}-H \ \xrightleftharpoons{\text{Add a Proton}} \ \left[H-\overset{..}{\underset{H}{N}}: \quad \underset{H_3C}{\overset{\cdot}{C}}=C\overset{CH_3}{\underset{H}{}} \right]$$

    - **Step 3:** One-electron reduction

        $$\left[H-\overset{..}{\underset{H}{N}}: \quad \underset{H_3C}{\overset{\cdot}{C}}=C\overset{CH_3}{\underset{H}{}} \right] + \cdot Na \ \xrightarrow{\text{One-Electron Reduction}} \ \left[\underset{H_3C}{\overset{\ominus}{\underset{..}{C}}}=C\overset{CH_3}{\underset{H}{}} \right]^{\oplus Na}$$

- **Step 4:** Add a proton

- Reaction Summary: Na transfers an electron to the alkyne producing a radical anion. The radical anion removes a proton from ammonia and a second atom of Na transfers another electron to the alkyne establishing the *trans* stereoselectivity. The carbanion removes a proton from ammonia to give a *trans*-alkene

  Alkyne —Na/NH$_{3(l)}$→ Alkene

  - Reaction starts with an alkyne
    - It is reduced to a *trans*-alkene by the end of the reaction
    - Reaction conditions do not reduce alkenes so the reaction does not proceed further than an alkene
  - Stereoselectivity: always produces a *trans* alkene with anti-addition
    - Hydrogens add to the opposite side
  - No carbocation rearrangement in this reaction

Reaction: Hydroboration – Oxidation of Alkynes

- Mechanism steps (using a terminal alkyne as an example)
    - **Step 1:** Simultaneous bond formation
        - This step requires (sia)$_2$BH:
        - In the figure above, you are looking at the transition state and not the intermediate
            - Dashed bond lines (not to be confused with dashed wedges) are showing bonds that are forming and bonds that are breaking
        - Boron adds to the carbon with less steric hindrance
        - Keep in mind that "R" is an abbreviation for any group or chemical chain in which a carbon or hydrogen atom is attached to the rest of the molecule
        - In the figure above, you are looking at the intermediate
    - **Step 2:** Chemist opens the flask and adds H$_2$O$_2$/NaOH
        - Result of this step: bond with boron is replaced with a hydroxyl group (-OH)

- Creates an enol
  - Enol – compound containing an OH group on one carbon of a C-C double bond

$$\left[\begin{array}{c} \text{H}_3\text{C} \\ \diagdown \\ \text{C}=\text{C} \\ \diagup \quad \diagdown \\ \text{H} \quad \text{H} \end{array} \begin{array}{c} \text{R} \\ \diagup \\ \text{B} \\ \diagdown \\ \text{R} \end{array} \right] \xrightleftharpoons[\text{(Chemist opens flask and adds new reagent)}]{\text{2. H}_2\text{O}_2 / \text{NaOH}} \left[\begin{array}{c} \text{H} \quad\quad \ddot{\text{O}}\text{H} \\ \diagdown \quad\quad \diagup \\ \text{C}=\text{C} \\ \diagup \quad\quad \diagdown \\ \text{H}_3\text{C} \quad\quad \text{H} \\ \text{Enol} \end{array} \right]$$

- **Step 3:** Keto-enol tautomerization
  - Enol is in equilibrium with a keto form through the displacement of a hydrogen from oxygen to carbon and through the migration of the double bond from C=C to C=O
    - Keto form predominates at equilibrium
    - Keto and enol forms are tautomers

$$\left[\begin{array}{c} \text{H} \quad\quad \ddot{\text{O}}\text{H} \\ \diagdown \quad\quad \diagup \\ \text{C}=\text{C} \\ \diagup \quad\quad \diagdown \\ \text{H}_3\text{C} \quad\quad \text{H} \\ \text{Enol} \end{array} \right] \xrightleftharpoons[\text{Keto-enol tautomerization}]{} \boxed{\begin{array}{c} \text{H} \quad\quad :\!\ddot{\text{O}}\!: \\ | \quad\quad || \\ \text{CH}_3-\text{C}-\text{C}-\text{H} \\ | \\ \text{H} \\ \text{Products} \end{array}}$$

- Reaction Summary: hydroboration of a terminal alkyne takes place using a sterically hindered (sia)$_2$BH to give a syn-product (an enol). Tautomerization of the enol gives an aldehyde.

$$\text{Terminal Alkyne} \xrightarrow[\text{2) H}_2\text{O}_2/\text{NaOH}]{\text{1) (sia)}_2\text{BH}} \text{Aldehyde}$$

$$\text{Internal Alkyne} \xrightarrow[\text{2) H}_2\text{O}_2/\text{NaOH}]{\text{1) BH}_3} \text{Ketone}$$

- Reaction starts with an alkyne
  - If the starting material is a terminal alkyne, it is transformed to an aldehyde by the end of the reaction
    - Terminal alkyne – alkyne with at least one hydrogen atom bonded to the triple bond

- If the starting material is an internal alkyne, it is transformed to an aldehyde by the end of the reaction
  - Internal alkyne – alkyne with no hydrogen atom bonded to the triple bond
- Regioselectivity: Non-Markovnikov
- Stereoselectivity: Syn-product
  - Hydroxyl group (-OH) and the hydrogen add to the same side
- No carbocation rearrangement in this reaction

Reaction: Electrophilic Addition to Alkynes

- In this reaction, an alkyne is hydrated to form an enol that tautomerises to form a ketone

$$\text{Internal Alkyne} \xrightarrow[H_2SO_4, H_2O]{HgSO_4} \text{Ketone}$$

- Mercury salt acts as a catalyst
  - Reaction without mercury is slow
- Regioselectivity: Markovnikov
- This reaction is very similar to the hydroboration reaction
  - However, note the different regioselectivity between the two reactions

Chapter 8: Haloalkanes and Radical Reactions

Haloalkanes (aka Alkyl Halide)

- Haloalkane (alkyl halide) - compound that contains a halogen atom (F, Cl, Br, etc.) covalently bonded to an sp^3 hybridized carbon
    - Typically represented by the symbol: RX
- Haloalkene (vinylic halide) – compound containing a halogen atom bonded to an sp^2 hybridized carbon
- Haloarene (aryl halide) – compound containing a halogen atom bonded to a benzene ring
    - Typically represented by the symbol: ArX
        - Ar = aromatic

Van der Waals Forces

- Haloalkenes are associated in the liquid state by van der Waals forces
    - Van der Waals forces pull molecules together
    - When molecules are brought close together, van der Waals attractive forces are overcome by repulsive forces between electron clouds of adjacent atoms or molecules
- Energy minimum - where the attractive and repulsive forces are equal
- Van der Waals radius - distance from the nucleus to the electron cloud surface

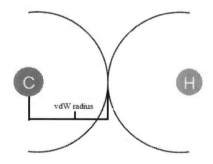

- When orbitals of two atoms are as close as possible but do not overlap, no bonds can be formed

- When atoms are close enough that their orbitals overlap, bonds can be formed

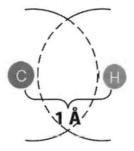

    - A covalent bond distance is about 1 angstrom (1 X10^{-10} meters)
    - A hydrogen bond distance is about 1.8 angstrom
- Nonbonded atoms in a molecule cannot approach each other closer than the sum of their van der Waals radii without causing nonbonded interaction strain

Boiling Points and Polarizability

- Haloalkanes typically have higher boiling points than alkanes (as long as they are of comparable size and shape)
    - Difference is due to the greater polarizability of the three unshared pairs of electrons on an halogen vs. the low polarizability of shared electron pairs of covalent bonds
        - Polarizability - ability of the electron cloud to distort or change in response to another molecule or ion
            - Larger volume occupied by the electron equates to a more polarized molecule
        - Polarizability ranking (weakest to strongest): F < Cl < Br < I
- Branched constitutional isomers have lower boiling points
    - Branching creates a more compact shape, decreased area of contact, decreased stacking, and decreased van der Waals interactions, which lower boiling points

Density

- Densities of liquid haloalkanes are greater than those of hydrocarbons of comparable molecular weight
- Halogens have a greater density than a methyl or methylene group
- Density ranking (least to most dense): CH_4 < CH_3Cl < CH_2Cl_2 < $CHCl_2$ < CCl_3

Bond Length and Strengths

- Bond length ranking (shorter to longer): C-H < C-F < C-Cl < C-Br < C-I
- Bond strength ranking (weakest to strongest): C-I < C-Br < C-Cl < C-H < C-F

Nomenclature

- Identify the parent chain (longest carbon chain) and number it (always number sequentially)
    - Numbering must also be done to give the substituent encountered first the lowest number, whether it is a halogen or an alkyl group
- Halogen substituents are indicated by prefixes: fluoro-, chloro-, bromo-, and iodo-
    - Must be listed alphabetically while naming
    - Must also indicate which carbon of the parent chain it is on by placing the carbon number preceding the name of the halogen with a hyphen
        - Example: 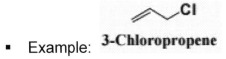 3-Chloropropene
- Priority order: double bonds > triple bonds > halogen substituent
    - Double bond must be given the lowest set of numbers possible

Radical Chain Mechanism

- Radical – chemical species that has one or more unpaired electrons
    - Radicals are formed by hemolytic bond cleavage
    - Order of stability of alkyl radicals: 3° > 2° > 1° > methyl

Steps of Chain Reaction

Chain length refers to the number of times the cycle of chain propagation steps repeats in a chain reaction.

- Chain initiation
    - Characterized by formation of reactive intermediates (radicals, anions, or cations)
- Chain propagation
    - Characterized by the reaction of a reactive intermediate and a molecule to form a new radical or reactive intermediate and a new molecule

- Chain termination
    - Characterized by the destruction of reactive intermediates

Reaction: Radical Halogenation of Alkanes

- Halogen reactivity: $F_2 > Cl_2 > Br_2 > I_2$
    - Only chlorination and bromination are useful in a laboratory setting
    - Bromination is more selective than chlorination
- Mechanism steps:
    - **Step 1:** Initiation
        - Heat or UV light causes homolytic bond cleavage
            - This type of bond breaking results in the bonding electron pair being split evenly between the products

    - **Step 2:** Propagation
        - End result is that a haloalkane is produced
            - Also, another Br radical is produced to propagate the reaction

- **Step 3:** Termination
  - Various reactions between the possible pairs of radicals occur
    - Form X₂, a hydrocarbon, and the product (a haloalkane)

$$:\!\ddot{B}r\cdot + \cdot\ddot{B}r\!: \rightleftharpoons\; :\!\ddot{B}r\!-\!\ddot{B}r\!:$$

[Diagrams showing ethyl radical + ethyl radical → butane, and ethyl radical + Br• → bromoethane]

- Reaction Summary: heat or UV light causes hemolytic cleavage of the weak halogen bond and generates two radicals. Bromine radical extracts a hydrogen to form H-Br and a methyl radical. The methyl radical extracts a bromine atom to form another bromine radical and the haloalkane product. Various reactions between radicals occur and they remove radicals from the reaction cycle.

$$\text{Alkane} \xrightarrow[\text{Br}_2 \text{ or Cl}_2]{h\nu \text{ or heat}} \text{Haloalkane}$$

  - Reaction starts with an alkane
    - It is transformed to a haloalkane by the end of the reaction
    - Reaction conditions do not reduce alkenes so the reaction does not proceed further than an alkene
    - Reaction proceeds through a radical chain mechanism
      - Involves radical intermediates
    - Termination steps have low probability
      - Radical species are present in low concentrations

Reaction: Allylic Bromination of Alkenes

- Allylic Carbon – C atom adjacent to a C-C bond
- Allylic Hydrogen – H atom on an allylic carbon

- Mechanism steps:
  - **Step 1:** Initiation
    - Heat or UV light causes homolytic bond cleavage

  - **Step 2:** Propagation
    - End result is that a bromoalkene is produced
      - Also, another Br radical is produced to propagate the reaction

  - **Step 3:** Termination
    - Various reactions between the possible pairs of radicals occur

- Reaction Summary:

$$\text{Alkene} \xrightarrow[\text{NBS}]{h\nu \text{ or heat}} \text{Bromoalkene}$$

  o Reaction starts with an alkene
    - It is transformed to a bromoalkene by the end of the reaction
    - Reaction proceeds through a radical chain mechanism
      - Involves radical intermediates
    - Termination steps have low probability
      - Radical species are present in low concentrations

Reaction: Radical Addition of HBr to Alkenes

- Addition of HBr to alkenes gives either Markovnikov addition or non-Markovnikov addition (depends on reaction conditions)
  o When radicals are absent, Markovnikov addition occurs
  o When peroxides or other sources of radicals are present, non-Markovnikov addition occurs
- Addition of HCl and HI gives only Markovnikov products
- Mechanism steps:
  o **Step 1:** Initiation

$$R-\ddot{O}-\ddot{O}-R \xrightleftharpoons{\text{heat}} R-\ddot{O}\cdot + \cdot\ddot{O}-R$$

$$H-\ddot{Br}: + [\cdot\ddot{O}-R] \rightleftharpoons [\cdot Br] + H-\ddot{O}-R$$

- **Step 2:** Propagation
  - End result is that a bromoalkane is produced
    - Also, another Br radical is produced to propagate the reaction

- **Step 3:** Termination
  - Various reactions between the possible pairs of radicals occur
  - Follows the same pattern discussed in the previous two reactions

- Reaction Summary:

$$\text{Alkane} \xrightarrow[\text{Br}_2]{\text{hv or heat}} \text{Bromoalkane}$$

  - Reaction starts with an alkane
    - It is transformed to a bromoalkane by the end of the reaction
    - Reaction proceeds through a radical chain mechanism
      - Involves radical intermediates
    - Termination steps have low probability
      - Radical species are present in low concentrations
  - Regioselectivity:
    - When radicals are absent, Markovnikov addition occurs
    - When peroxides or other sources of radicals are present, non-Markovnikov addition occurs
  - Carbocation rearrangement is possible in this reaction

Chapter 9: Nucleophilic Substitution and β-Elimination

Nucleophilic Substitution

Nucleophilic substitution reactions are any reactions in which one nucleophile substitutes for another at a tetravalent carbon.

- Nucleophile – molecule or ion that donates a pair of electrons to another atom or ion to form a new covalent bond
    - Nucleophiles are also Lewis bases
- There are two limiting mechanisms for nucleophilic substitutions
    - Fundamental difference between the two mechanisms is the timing of bond-breaking and bond-forming steps

S_N2 Reactions

- S_N2 reaction – reaction during which bond-breaking and bond-forming steps occur simultaneously (simultaneous attack of the nucleophile and departure of the leaving group)
    - S = substitution
    - N = nucleophilic
    - 2 = bimolecular (both reactants are involved in the rate-determining step)

Transition state with simultaneous bond breaking and bond forming

- S_N2 reactions result in inversion at the chiral center

Kinetics of S$_N$2 Reactions

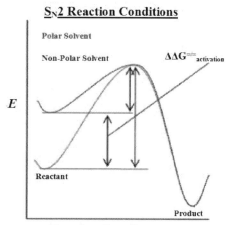

S$_N$2 Reaction Energy Diagram:

- Reaction occurs in one step
- Reaction leading to the transition state involves both the haloalkane and the nucleophile
    - Results in a 2nd order reaction

S$_N$1 Reactions

- S$_N$1 reaction – reaction during which the bond-breaking between carbon and the leaving group is entirely completed before bond-forming with the nucleophile begins
    - S = substitution
    - N = nucleophilic
    - 1 = unimolecular (only one reactant is involved in the rate-determining step)
- In an S$_N$1 reaction, the R and S enantiomers are formed in equal amounts
    - The product is a racemic mixture
- Rearrangements are common in S$_N$1 reactions
    - Make sure to check to see if the initial carbocation can rearrange to a more stable one

Kinetics of S$_N$1 Reactions

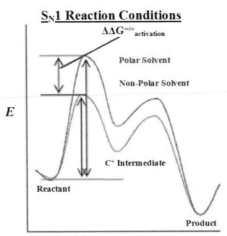

S$_N$1 Reaction Energy Diagram:

- Reaction occurs in two steps
- Reaction leading to formation of the transition state for the carbocation intermediate involves only the haloalkane and not the nucleophile
    - Result in a 1st order reaction

Leaving Groups

- More stable an anion, the better its leaving ability
    - Most stable anions are conjugate bases of strong acids
- Ranking of reactivity as a leaving group (most reactive to least reactive): I$^-$ > Br$^-$ > Cl$^-$ > H$_2$O >> F$^-$ > CH$_3$COO$^-$ > HO$^-$ > CH$_3$O$^-$ > NH$_2^-$

Solvents

- Protic solvent - solvent that is a hydrogen bond donor
    - Most common protic solvents contain -OH groups
- Aprotic solvent - solvent that cannot serve as a hydrogen bond donor
    - Nowhere in the molecule is there a hydrogen bonded to an atom of high electronegativity

- Solvents are classified as polar and nonpolar
    - Most common measure of solvent polarity is the dielectric constant
        - Dielectric constant - measure of a solvent's ability to insulate opposite charges from one another
            - Greater value of the dielectric constant of a solvent equates to a smaller interaction between ions of opposite charge dissolved in that solvent
            - Polar solvents have a dielectric constant > 15
            - Nonpolar solvents have a dielectric constant < 15

Solvent – S_N2

- Most common type of S_N2 reaction involves a negative nucleophile and a negative leaving group
    - The weaker the solvation of a nucleophile, the less the energy required to remove it from its solvation shell and the greater the rate of S_N2

Solvent – S_N1

- S_N1 reactions involve creation and separation of unlike charges in the transition state of the rate-determining step
- Rate depends on the ability of the solvent to keep these charges separated and to solvate both the anion and the cation
- Polar protic solvents are the most effective solvents for S_N1 reactions

Nucleophilicity and Basicity

Since all nucleophiles are also bases, the correlation between nucleophilicity and basicity are studied.

- Nucleophilicity - kinetic property measured by the rate at which a nucleophile causes a nucleophilic substitution under a standardized set of experimental conditions
- Basicity - equilibrium property measured by the position of equilibrium in an acid-base reaction

- Polar protic solvents
    - Anions are highly solvated by hydrogen bonding with the solvent
    - The more concentrated the negative charge of the anion, the more tightly it is held in a solvent shell
        - The nucleophile must be at least partially removed from its solvent shell in order to participate in S_N2 reactions
        - Nucleophilicity ranking: $I^- > Br^- > Cl^- > F^-$
- Polar aprotic solvents are very effective in solvating cations
    - But they are not nearly so effective in solvating anions
        - This means they participate readily in S_N2 reactions
        - Nucleophilicity parallels basicity: $F^- > Cl^- > Br^- > I^-$
- In general, in a series of reagents with the same nucleophilic atom: anionic reagents are stronger nucleophiles than neutral reagents
    - This trend parallels the basicity of the nucleophile
- In general, when comparing groups of reagents in which the nucleophilic atom is the same: the stronger the base, the greater the nucleophilicity

Summary of S_N2 and S_N1

Type of Alkyl Halide	S_N2	S_N1
Methyl (CH_3X)	S_N2 is favored.	S_N1 does not occur.
Primary (RCH_2X)	S_N2 is favored.	S_N1 rarely occurs.
Secondary (R_2CHX)	S_N2 is favored in aprotic solvents with good nucleophiles.	S_N1 is favored in protic solvents with poor nucleophiles. Look out for carbocation rearrangement.
Tertiary (R_3CX)	S_N2 does not occur because of steric hindrance around the reaction center.	S_N1 is favored.
Substitution at a Stereocenter	Inversion of configuration. The nucleophile attacks the stereocenter from the side opposite the leaving group.	Racemization is favored. Carbocation intermediate is planar, and attack of the nucleophile occurs with equal probability from either side.

β-Elimination

β-Elimination is a type of reaction in which a small molecule (e.g., HCl, HBr, HI, or HOH) is split out or eliminated from adjacent carbons.

- Zaitsev Rule
    - Major product of a β-elimination is the more stable (more substituted) alkene
    - Example:

        $$\text{2-Bromo-2-methylbutane} \xrightarrow[\text{CH}_3\text{CH}_2\text{OH}]{\text{CH}_3\text{CH}_2\text{O}^-\text{Na}^+} \text{2-Methyl-2-butene (major product)} + \text{2-Methyl-1-butene}$$

- There are two limiting mechanisms for β-elimination reactions (E1 and E2)

E2 Mechanism

- Breaking of the R-LG and C-H bonds occurs simultaneously
    - Both R-LG and the base involved in breaking of the C-H bond are involved in the rate-determining step
- Kinetics of E2
    - Reaction occurs in one step
    - Reaction is 2nd order
- Regioselectivity
    - With a strong base, the major product is the more stable (more substituted) alkene
    - With a strong and sterically hindered base such as tert-butoxide, the major product is often the less stable (less substituted) alkene

E1 Mechanism

- Breaking of the R-LG bond to give a carbocation is fully completed before reaction with the base to break the C-H bond
    - Only breaking of the R-LG bond is involved in the rate-determining step
- Kinetics of E1
    - Reaction occurs in two steps
    - Rate-determining step is carbocation formation
    - Reaction is 1st order
- Regioselectivity
    - E1 always follows the Zaitsev rule
        - Major product is the more stable (more substituted) alkene

Summary of E2 vs. E1

Type of Alkyl Halide	E2	E1
Primary (RCH$_2$X)	E2 is favored.	E1 does not occur.
Secondary (R$_2$CHX)	Main reaction with strong bases such as OH and OR.	Main reaction with weak bases such as H$_2$O, ROH.
Tertiary (R$_3$CX)	Main reaction with strong bases such as OH and OR.	Main reaction with weak bases such as H$_2$O, ROH.

S_N vs. E

- Because many nucleophiles are also strong bases (e.g., -OH and -OR) S_N and E reactions often compete

- Ratio of S_N/E products depends on the relative rates of the two reactions

Type of Alkyl Halide	Favored Reaction(s)	How to Decide Which Reaction is Occurring
Methyl (CH$_3$X)	Only S$_N$2	
Primary (RCH$_2$X)	S$_N$2 or E2	Strong, bulky base such as (t-BuOK) → E2 No strong, bulky base → S$_N$2
Secondary (R$_2$CHX)	S$_N$2, E2, and S$_N$1/E1 (Occur Together)	Strong base → E2 Polar protic solvent and no strong base → S$_N$1/E1 Strong nucleophile and no polar protic solvent or strong base → S$_N$2
Tertiary (R$_3$CX)	E2 or S$_N$1/E1 (Occur Together)	Strong base → E2 No strong base → S$_N$1/E1

Chapter 10: Alcohols and their Reactions

Alcohols

Structure

- Functional group of an alcohol is an -OH group bonded to an sp^3 hybridized carbon
    - Bond angles = ~109.5 about the hydroxyl oxygen atom
- Oxygen is sp^3 hybridized
    - Two sp^3 hybrid orbitals form sigma bonds to a carbon and hydrogen
    - The remaining two sp^3 hybrid orbitals each contain an unshared pair of electrons

Nomenclature

- Parent chain is the longest chain that contains the OH group
- Priority order: hydroxyl group > double bonds > triple bonds > halogen substituent
    - Number the parent chain to give the OH group the lowest possible number
- Change the suffix –e to ol

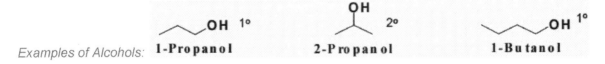

Examples of Alcohols: 1-Propanol   2-Propanol   1-Butanol

- Compounds containing more than one OH group are named diols, triols, etc.

Physical Properties

- Alcohols are polar
    - Interact with themselves and with other polar compounds by dipole-dipole interactions
        - Dipole-dipole interaction – attraction between positive end of one dipole with the negative end of another

- Hydrogen bonding occurs when the positive end hydrogen bonding of one dipole is an H bonded to F, O, or N (atoms of high electronegativity) and the other end is F, O, or N
    - Hydrogen bonds are weaker than covalent bonds but still contribute to physical properties
- Molecules with higher hydrogen bonding interactions typically have a higher boiling point
- Relative to alkanes (of comparable size and molecular weight)
    - Alcohols have higher boiling points and are more soluble in water
        - Additional –OH groups increase solubility and boiling points

Reaction: S$_N$2 Reaction of 1° Alcohol with HX

- Mechanism steps
    - **Step 1:** Add a proton

    - **Step 2:** S$_N$2 (Simultaneous Attack of the Nucleophile and Departure of the Leaving group)

- Reaction Summary: first step of the reaction is the protonation of the OH group to convert it to a good leaving group. This is the fast step. The second step is simultaneous attack of the nucleophile and departure of the leaving group. This is the rate determining step. Primary alcohols produce a haloalkane with HX by S$_N$2 mechanism.

$$\text{Primary Alcohol} \xrightarrow{HX} \text{Haloalkane}$$

  o Reaction starts with a primary alcohol
    - Transformed to an haloalkane by the end of the reaction

Reaction: S$_N$1 Reaction of 2° and 3° Alcohol with HX

- Mechanism steps
  o **Step 1:** Add a proton
  o **Step 2:** S$_N$1 – Departure of the Leaving Group
  o **Step 3:** S$_N$1 – Attack of the Nucleophile

- Reaction Summary:

$$\text{2° or 3° Alcohol} \xrightarrow{\text{HX}} \text{Haloalkane}$$

  o Reaction starts with a secondary or alcohol
    - Transformed to an haloalkane by the end of the reaction
  o When this reaction occurs with a secondary alcohol, carbocation rearrangement is possible due to hydrogen shift

Reaction: S$_N$2 Reaction of 1° and 2° Alcohol with PBr$_3$

- This is an alternative method for the synthesis of 1° and 2° bromoalkanes
  o Results in less rearrangement than reaction with HX
- Mechanism steps
  o **Step 1:** Attack of the nucleophile

  o **Step 2:** S$_N$2 (Simultaneous Attack of the Nucleophile and Departure of the Leaving group)

- Reaction Summary:

$$\text{1° or 2° Alcohol} \xrightarrow{\text{PBr}_3} \text{Haloalkane}$$

    o Reaction starts with a primary or secondary alcohol
        ▪ Transformed to an haloalkane by the end of the reaction
    o Steroselectivity: Reaction results in inversion at the chiral center

Reaction of Alcohols with Thionyl Chloride (SOCL$_2$)

- Mechanism steps
    o **Step 1:** Attack of the nucleophile

    o **Step 2:** Remove a proton

    o **Step 3:** S$_N$2 (Simultaneous Attack of the Nucleophile and Departure of the Leaving group)

- Reaction Summary:

$$\text{Alcohol} \xrightarrow[\text{Pyridine}]{\text{SOCl}_2} \text{Alkyl Chloride}$$

  o Reaction starts with a alcohol
    ▪ Transformed to an alkyl chloride by the end of the reaction
  o Steroselectivity: Reaction results in inversion at the chiral center

Reaction: Dehydration of 2° and 3° Alcohol

- Mechanism steps
  o **Step 1:** Add a proton
    ▪ First, H_2SO_4 reacts with H_2O to produce H_3O^+ and HSO_4^-

  o **Step 2:** E1 – Departure of the leaving group

  o **Step 3:** Remove a proton

- Reaction Summary:

$$\text{2° or 3° Alcohol} \xrightarrow{\text{HX}} \text{Alkene}$$

  - Reaction starts with a secondary or tertiary alcohol
    - Transformed to an alkene by the end of the reaction
  - Regioselectivity: Zaitsev product
    - Alkene with the greater number of substituents on the double bond predominates
  - Steroselectivity: E-product
    - Higher priority groups are on the opposite side

Reaction: Chromic Acid Oxidation of Alcohols

- Mechanism steps
  - **Step 1:** Several steps
    - Typically, you aren't going to be required to know the exact steps that take place, but you are probably going to be asked to remember the product of the first step

  - **Step 2:** Remove a proton

  - **Step 3:** Add a proton

- Step 4: Attack of the nucleophile

- Step 5: Remove a proton

- Step 6: Several Steps

- Step 7: Remove a proton

• Reaction Summary:

Primary Alcohol $\xrightarrow{H_2CrO_4}$ Aldehyde $\xrightarrow{H_2CrO_4}$ Carboxylic Acid

2° Alcohol $\xrightarrow{H_2CrO_4}$ Ketone

- Reaction that starts with a primary alcohol, results in the formation of a aldehyde or carboxylic acid depending on the reaction conditions

- Reaction that starts with a secondary alcohol, results in the formation of a ketone

- Tertiary alcohols are not oxidized by chromic acid (H_2CrO_4)

PCC Oxidation of Alcohols

- Pyridinum chlorochromate (PCC) is selective for the oxidation of 1° alcohols to aldehydes
    - It doesn't oxidize aldehydes further to carboxylic acids
- PCC oxidizes a 2° alcohol to a ketone

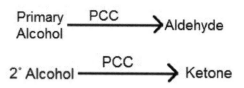

Chapter 11: Ethers and Epoxides

Nomenclature

- Parent chain is the longest carbon chain in a molecule

- Name the OR group as an alkoxy substituent

	Suffix	Prefix
-OH	-ol	Hydroxy-
-C=C-	-ene	
-C≡C-	-yne	
X, R, OR		Halo, alkyl, alkoxy

Cyclic Ethers

- Prefix ox- is used to represent an oxygen that is part of a ring structure

- Suffixes –irane, -ethane, -olane, and –ane are used to indicate whether three, four, five, and six atoms are in a saturated ring

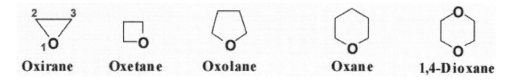

Oxirane   Oxetane   Oxolane   Oxane   1,4-Dioxane

Physical Properties of Ethers

- Ethers are polar compounds

    - Despite being polar compounds, only weak dipole-dipole attractive forces exist between their molecules in the pure liquid state

- Boiling points of ethers are lower than alcohols (as long as they are of comparable molecular weights)

- Boiling points of ethers are close to those of hydrocarbons (as long as they are of comparable molecular weights)

- Ethers are hydrogen bond acceptors

Preparation of Ethers

Williamson Ether Synthesis

- Synthesis of ethers by the S_N2 displacement of halide, tosylate, or mesylate by alkoxide ion

General Williamson Ether Synthesis Reaction:

$$R-O^{\ominus} + R'-X \xrightarrow{(S_N2)} R-O-R' + X^-$$

- Reaction works best (i.e. yields are highest) with methyl and 1° halides
    - Yields are lower with 2° halides
        - Yields are lower because of competing β-elimination reaction that occurs simultaneously
    - Reaction fails with 3° halides
        - Only β-elimination reaction occurs, not S_N2

Acid-catalyzed Dehydration of Alcohols

- This is a specific example of an S_N2 reaction, where a poor leaving group (OH⁻) is converted to a better one (H_2O)

$$R-\boxed{OH + H}-OR' \xrightarrow[\text{Heat}]{H_2SO_4} R-OR'$$

Acid-catalyzed Addition of Alcohols to Alkenes

$$\underset{R}{\overset{R}{>}}C=CH_2 + H-OR' \xrightarrow[\text{Heat}]{H_2SO_4} \underset{R}{\overset{R}{>}}C\underset{CH_2H}{\overset{OR'}{<}}$$

- Highest yields when using an alkene that can form a stable carbocation
    - Or, using a 1° alcohol that is not likely to undergo acid-catalyzed dehydration

Synthesis of Epoxides

Epoxides are cyclic ethers in which one of the atoms of a 3-membered ring is oxygen.

Example of an Epoxide (Oxirane):

- Simple epoxides are named as derivatives of oxirane
- Prefix epoxy- is used to represent when the epoxide is part of another ring system

Air Oxidation of Ethylene

$$2\ H_2C=CH_2 \xrightarrow[Ag]{O_2} 2\ H_2C\underset{O}{\triangle}CH_2$$

- Typically seen manufactured using this method for use in industry

Epoxide Formation with Peroxycarboxylic Acid

$$\underset{R}{\overset{R}{\diagdown}}C=CH_2 \xrightarrow[CH_2Cl_2]{RCO_3H} \text{epoxide}$$

- In this reaction, an alkene reacts with peroxycarboxylic acid in a single step (electrophilic addition) to produce an epoxide
- Stereoselectivity: Diastereoselective
  - Stereoisomer you get depends on the configuration of the alkene you start with
    - A *cis*-2-butene gives only *cis*-2,3- dimethyloxirane
    - A *trans*-2-butene gives only *trans*-2,3-dimethyloxirane

Preparation of Epoxide from a Halohydrin

$$CH_3CH=CH_2 \xrightarrow{Cl_2,\ H_2O} CH_3\underset{Cl}{\overset{OH}{CH}}-CH_2 \xrightarrow[S_N2]{NaOH,\ H_2O} CH_3CH\underset{O}{\triangle}CH_2$$

- Stereoselectivity: Diastereoselective
  - Stereoisomer you get depends on the configuration of the alkene you start with

cis-2-Butene → (1. Cl_2, H_2O; 2. NaOH, H_2O) → *cis*-2,3-Dimethyloxirane

Ring Opening of Epoxides

3-membered rings have strains associated with them, because of this; they readily undergo a variety of ring-opening reactions.

Acid Catalyzed Hydrolysis of Epoxides

- Mechanism steps
    - **Step 1:** Add a proton

    - **Step 2:** Attack of the nucleophile

    - **Step 3:** Remove a proton

- Reaction Summary: in acid, the epoxide oxygen is protonated to make a bridged oxonium ion. The nucleophile (H_2O) attacks the carbon which is more carbocation-like from the opposite side. Removing a proton gives a diol.

$$\text{Epoxide} \xrightarrow[H_2O]{H_2SO_4} \text{Diol}$$

- Regioselectivity: nucleophile attacks the most substituted Carbon
- Stereoselectivity: Anti-product
    - -OH groups are on opposite sides

Reaction: Base Catalyzed Hydrolysis of Epoxides

- Mechanism steps

    - **Step 1:** S$_N$2 (Simultaneous Attack of the Nucleophile and Departure of the Leaving Group)

    - **Step 2:** Remove a proton

- Reaction Summary:

    $$\text{Epoxide} \xrightarrow[\text{H}_2\text{O}]{\text{OH}^-} \text{Diol}$$

    - Regioselectivity: nucleophile attacks the least substituted Carbon
    - Stereoselectivity: Anti-product
        - -OH groups are on opposite sides

CONCLUDING REMARKS

I hope this book has provided you tremendous value for your money and has helped you do better on your exams! If it has done both of these things, I have achieved my purpose in making this guide.

Furthermore, my goal is to create more books and guides that continue to deliver great value to readers like you for little monetary costs. Thank you again for purchasing this study guide and I wish you the best on your future endeavors!

- Dr. Holden Hemsworth

More Books By Holden Hemsworth

Do You Need Help with Other Classes?

Check out Other Books in the Ace! Series

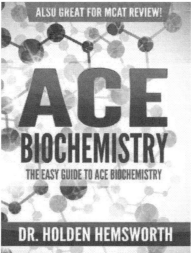

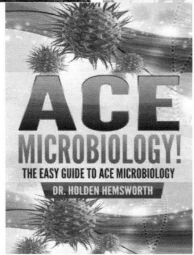
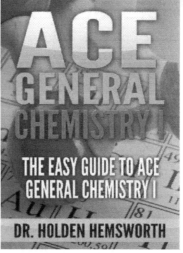

More Books Coming Soon!

Made in the USA
Middletown, DE
27 August 2019